'Women Entrepreneurs: A Study of Achievement Motivation, Leadership and Work Life Balance of Women Entrepreneurs of Gujarat'

Sharma, Jessica Maria

INDEX

Sr. no	Particulars	Page no.
*	Certificate	II
*	Preface	III
*	Acknowledgement	IV
*	List of Tables	VIII
*	List of Figures	XIII
1	Introduction	1
2	Review of Literature	27
3	Research Methodology	53
4	Data Analysis and Interpretation	60
5	Findings, Discussion and Conclusion	150
6	Suggestions and Recommendations	169
7	Bibliography	185
	Annexure	

LIST OF TABLES

Table no.	Particulars	Page no.
1.1	Showing Age of the Respondents	62
1.2	Showing Educational Qualifications of the Respondents	64
1.3	Showing the Marital Status of Respondents	65
1.4	Showing the experience of respondents (in years)	66
1.5	Showing Experience of running current enterprise	67
1.6	Showing average monthly income of the respondents	68
1.7	Showing husband's occupation	69
1.8	Showing Husband's Monthly Income	70
1.9	Showing Husband's Educational Qualification	72
1.10	Showing number of Family Members	74
1.11	Showing number of children of the respondents	75
2.1	Showing Achievement Motivation of Respondents	76
2.1.1	Cumulative response on Achievement Motivation	80
3.1	Showing Responses of Respondents on Leadership	82
3.1.1	Showing Cumulative responses on	86

	Leadership	
4.1	Showing Work Life Balance of Respondents	87
4.1.1	Showing cumulative responses of Work Life Balance of Respondents	93
5.1	Showing Association between Age and Achievement Motivation	95
5.2	Showing Association between Education and Achievement Motivation	96
5.3	Showing Association between Experience and Achievement Motivation	97
5.4	Showing Association between Average Income (Per Month) and Achievement Motivation	98
5.5	Showing Association between Age and Leadership	100
5.6	Showing Association between Education and Achievement Motivation	101
5.7	Showing Association between Experience and Leadership	102
5.8	Showing Association between Income and Leadership	103
5.9	Showing Association between Age and Work Life Balance	104
5.10	Showing Association between Education and Work Life Balance	105
5.11	Showing Association between Experience and Work Life Balance	106

5.12	Showing Association between Income and Work Life Balance	108
5.13	Showing Association and Association between Independent and Dependent Variables.	110
6.1	Showing correlation between Age and Achievement Motivation, Leadership and Work life balance	112
6.1.1	Showing Multiple Comparisons of Age and Achievement Motivation	114
6.2	Showing correlation between Educational Qualification and Achievement Motivation, Leadership and Work Life Balance	115
6.3	Showing correlation between Marital Status and Achievement Motivation, Leadership and Work Life Balance	116
6.4	Showing correlation between Number of Years of Experience and Achievement Motivation, Leadership and Work Life Balance	117
6.5	Showing correlation between Experience of running own enterprise and Achievement Motivation, Leadership and Work Life Balance	118
6.5.1	Showing Multiple Comparisons among Groups for Variance	120
6.6	Showing correlation between Average Monthly Income and Achievement	121

	Motivation, Leadership and Work Life Balance	
6.6.1	Showing Multiple Comparisons for Average Income (Per Month) and Leadership and Work life Balance	123
6.7	Showing correlation between Spouse's Occupation and Achievement Motivation, Leadership and Work Life Balance	124
6.7.1	Showing Multiple Comparisons of Husband's Occupation and Achievement Motivation	125
6.8	Showing correlation between Spouse's Average Monthly Income and Achievement Motivation, Leadership and Work Life Balance	126
6.8.1	Showing multiple comparisons between Husband's Income (Per Month) and Work life Balance	127
6.9	Showing correlation between Number of Family Members and Achievement Motivation, Leadership and Work Life Balance	128
6.9.1	Showing multiple comparisons of No. of Family Members and Leadership	130
6.10	Showing correlation between Spouse's Educational Qualification and Achievement Motivation, Leadership and Work Life Balance	131
6.10.1	Showing multiple comparisons between Husband's Educational Qualifications and	132

	Achievement Motivation	
6.11	Showing correlation between Number of Children of respondents and Achievement Motivation, Leadership and Work Life Balance	133
6.12	Showing correlation between Average Family Income (Per Month) and Achievement Motivation, Leadership and Work Life Balance	134
6.12.1	Showing multiple comparisons between Income of Family (Per Month) and Leadership and Work life Balance	136
6.4.1	Showing Acceptance/ Rejection of Hypotheses	163

LIST OF FIGURES

Figure	Particulars	Page no.
1.1	Showing Age of Respondents	62
1.2	Showing Educational Qualification of respondents	63
1.3	Showing marital status of respondents	64
1.4	Showing total years of experience of respondents	65
1.5	Showing respondents' experience of running own enterprise	66
1.6	Showing Average Monthly Income of Respondents	67
1.7	Showing Occupation of Respondents' Husbands	68
1.8	Showing Average Monthly Income of Respondents' Husbands	70
1.9	Showing Respondent's Husband's Educational Qualification	71
1.10	Showing Number of Family Members of Respondents	72
1.11	Showing Number of Children of Respondents	73
2.1.1	Showing Achievement Motivation of Respondents	77
3.1.1	Showing Cumulative responses on Leadership	82
4.1.1	Showing cumulative responses of Work life Balance of Respondents	87

PREFACE

We live in a society which is heavily conditioned with patriarchal school of thought. The gendered lens gives an absolutely sexist approach to everything that people do and/or say. People categorize every element to either be masculine or feminine; be it work, play or any other chore, it has been categorised into the gender category. This is the reason why working outside the house is considered masculine while working in the kitchen is considered feminine. In such categorizations, the essence of the word 'work' is lost. Work, per se, is beyond the categories of gender. Work is work, no matter who does it. But our conditioned thoughts have made us to categorise work in genders.

From triumphant home-makers to protectors of the nation, we find women in all spheres. Success stories of women who have made a difference to the society are widely published and publicised. These victorious achievements of women led to the undertaking of this study.

Women are considered to be vulnerable but this vulnerability is also their strength. It is often quoted that struggles make a person stronger. Same is the case when it comes to women. Women have always borne the brunt of the social dogma. Hence, they were seldom encouraged to take up anything that was not mainstream. But today, women have phenomenally excelled in all fields.

Female Entrepreneurship is a fascinating area of research wherein researchers are constantly trying to find the factors which make up an enterprising female - be it family-owned and inherited or start-up, women are proving to be good at managing it all. It therefore is interesting to know the factors which motivate women and enhance their leadership and also allows them to balance their personal life and work-life by giving both the aspects equal time and consideration.

CHAPTER I

INTRODUCTION

"A single person can move mountains, perform miracles.

And that person could be you..."

— Rashmi Bansal

There are a number of schools of thought, which view the idea of entrepreneurship from essentially different perspectives. The term has been used to define a wide range of activities such as creation, funding, adapting, and managing and venture.

According to P. F. Drucker an entrepreneur is one who constantly searches for change, answers to it and exploits it as an opportunity."

According to Richard Cantillion "entrepreneur is the agent who buys means of production at certain prices, in order to sell at prices that are certain at the moment at which he/she commits himself/ herself to his/ her cost".

The word ' entrepreneur' originates from the French word "Entreprendre" (to undertake) in the early 16th Century it was applied to persons involved in military expeditions, and extend to cover construction and civil engineering activities in the 17th century, but during the 18thcentury, the word 'entrepreneur' was used to refer to economic activities. Many authors have defined 'entrepreneur' differently. In general, an entrepreneur is a person who combines capital and labour for production. Entrepreneurs are the real heroes of economic life. They are the ones who sweat and toil and sacrifice and that natural resource gain value only by the ingenuity and labour of man. From the knowledge of failure, they forge success. In accepting risk, they achieve security for all. In embracing change, they make sure social and economic stability.

Entrepreneurship is neither a science, nor an art. It is a practice. However, as in all practices, medicine or engineering for instance, knowledge in entrepreneurship is a means to an end. It is not a 'flash of genius' but purposeful tasks that can be organized into systematic work.

The term entrepreneurship appears to connote a much-restricted meaning in the Indian context. It covers only a limited sphere of enterprising endeavour i.e., establishing and running of factories and industrial enterprises alone. It has also been viewed as a phenomenon occurring around an individual and benefitting only an individual. It is only recently that it is being appreciated as one that could be harnessed to benefit larger groups.

Women make up around half of the total world population. So is in India also. They are therefore, considered as the better half of the society. In traditional societies, they were confined to the four walls of houses performing household activities. The traditional set up is changing in the modern times. The change of social fabric of the Indian society, in terms of improved educational status of women and varied ambitions for better living, necessitated a transformation in the life style of Indian women. (Kumari, 2012)

Women, in the Indian society are conditioned to be extremely adaptive to the changes around. They are made to sit on the highest pedestal by making them feel that they uphold their family's integrity. Ironically, these women who uphold the integrity of their families are least respected and mostly humiliated. They are conditioned to be a doormat on which their husbands can always trudge. They are conditioned to respect their husband and his family even at the cost of her own self-respect. Self-respect, again, is a gendered concept. Men must have it while women can do without it. Many sexist practices exist in our country, which push women to the lowest rung of the ladder of empowerment. If at all women fight for their rights seeking equal opportunities, they are perceived in a different light. Each older generation of women compares itself to the newer generation and always make the new generation feel guilty of having asserted themselves for something, which should have been naturally given to them.

Being a culturally rich country, the people of India revere a lot of goddesses but on the other hand despise to give equal rights and opportunities to its women. Therefore, women need to put additional efforts to attain their fair share of opportunities in an otherwise patriarchal Indian society.

In the Indian society, a woman's role was naturally limited to the family. Since she was the bearer of children, she was fully occupied with her duties as a mother and a homemaker. It was societal expectation that women need to be bound to their homely duties and take care of the household, children and family. Women were restricted to experimenting only in the kitchens and were seldom encouraged to exhibit their skills of innovation outside their homes.

Many factors like urbanization, technical progress, woman's education, etc., have profoundly changed these traditional conditions even in a developing country like India. Women perceive more and more clearly that if they really want to contribute to the welfare of their family and society the most effective way is to go out of home and earn money.

The participation of women in the economic development process can be mainly categorised into four segments namely

- Employment in unorganized sector
- Employment in organized sector
- Self-employment and
- Entrepreneurs

Recently a new trend has emerged in India wherein women are venturing as entrepreneurs and are contributing to the economic development of the country. Women are increasingly seeking entrepreneurship as an avenue for economic growth. The government is also playing a significant role in mobilising women to become entrepreneurs through its different programs. This is helping is the number of women entrepreneurs increase much more.

The all round development of women has been one of the principal points of planning process in India. The First Five-Year Plan (1951-56) envisioned a number of welfare measures for women. Establishment of the Central Social Welfare Board, organization of Mahila Mandals and the Community Development Programmes were a few steps in this direction.

In the second Five-Year Plan (1956-61), the empowerment of women was closely related with the overall approach of intensive agricultural development programmes.

The Third and Fourth Five-Year Plans (1961-66 and 1969-74) supported female education as a leading welfare measure.

The Fifth Five-Year Plan (1974-79) highlighted training of women, who were in need of income and protection. This plan corresponded with International Women's Decade and the submission of Report of the Committee on the Status of Women in India. In 1976, Women's welfare and Development Bureau was started under the Ministry of Social Welfare.

The Sixth Five-Year Plan (1980-85) saw a certain change from welfare to development. It acknowledged women's lack of access to resources as a serious factor impending their growth.

The Seventh Five-Year Plan (1985-90) highlighted the need for gender equality and empowerment. For the first time, importance was placed upon qualitative aspects such as inculcation of confidence, generation of awareness with regards, to rights and training in skills for better employment.

The Eight Five-Year Plan (1992-97) focused on empowering women, especially at the gross roots level, through Panchayat Raj Institutions.

The Ninth Five-Year Plan (1997-2002) implemented a strategy of women's component plan, under which not less than 30 percent of funds/benefits were set aside for women-specific programmes.

The Tenth Five-Year Plan (2002-07) aims at empowering women through converting the recently adopted National Policy for Empowerment of Women (2001) into action and confirming Survival, Protection and Development of women and children through rights based approach. (Rao, 2007)

The Working Group on Empowerment of Women for the Eleventh Five Year Plan (2007-12), constituted by the Planning Commission in 2006, has emphasized the strengthening of Self-help Groups (SHGs) and community-based organizations for the empowerment of women. (Rajarajeswari, 2012)

The Twelfth Five Year Plan (2012-2016) aims at approving that RMK should be restructured from Society into Non-Deposit Taking Non-Banking-Finance-Company (ND-NBFC) to enable it to play its mandated role of women's empowerment through micro-finance and to extend capacity building and other support to women SHGs as a single window facility.

Total number of establishments owned by women entrepreneurs	8,050,819
Rural establishments owned by women entrepreneurs	5,243,044(65.12Percent)
Urban establishments owned by women entrepreneurs	2,807,775(34.88Percent)
Women owned establishments without hired workers	6,697354 (83.19Percent)
Women owned establishments with hired workers	1,35,3465(16.31Percent)
Women owned agricultural establishments	2.76 Million (34.3Percent)
Women owned non agricultural establishments	5.29 Million (65.7Percent)

Source: All India report of Sixth economic census 2016.

(Table showing overview of Women Entrepreneurs of India)

Women work participation is low in India in comparison to certain countries of the world. Women work participation in India is 31.6 % whereas in USA it is 45 %, UK 43 %, Canada 42 %, France 32 %, Indonesia 40 %, Sri Lanka and Brazil both 35 %. (Sangolagi & Alagawadi, 2016) In India, men generally take the lead in the entrepreneurial world. With the varying cultural norms and rise in literacy rate more and more Indian women are accepting entrepreneurship as their career. With the arrival of media, women are being aware of their own traits, rights and also the work situation. They are thriving as designers, interior decorators, exporters, publishers, garment manufacturers and still exploring new avenues of economic participation.

HISTORICAL EVOLUTION OF WOMEN ENTREPRENEURS

Up through the nineteenth century, women-owned businesses primarily included taverns and alehouses, millinery and retail shops, and brothels, and were often operated as a way to provide an income for women who found themselves without a breadwinning man. Sometimes, too, women inherited businesses from fathers or husbands and found themselves thrust into traditionally male enterprises. These women owned businesses in the face of prevailing social ideas that the unsavoury world of business was unsuitable for women's gentle and frail natures. Business, then, in these earlier eras, was a way for a woman in potentially dire circumstances to provide for herself rather than become a social burden. (From Ideas to Independence: A Century of Entrepreneurial Women, n.d.) Many women succeeded in any case; among them, Rebecca Lukens, who, beginning in 1825, turned her family's faltering ironworks into a thriving steel business that endured into the 21st century. From 1900 through 1929, Progressivism, feminism, consumerism and immigration all gave rise to a climate that was not only conducive to women's entrepreneurship, but also highly accepting of them.

World War II brought many women into the workforce, filling jobs so men could go off and fight. That same patriotic fervour also inspired many women to consider starting businesses of their own. When World War II ended, women were pushed from wartime jobs for returning soldiers, and many went straight into businesses of their own. For women whose husbands returned injured or did not return at all, there was an added incentive. Hopeful women entrepreneurs found plenty of help and encouragement. (From Ideas to Independence: A Century of Entrepreneurial Women, n.d.)

The Civil Rights and women's movements of the 1960s and 1970s brought a new sense of purpose and a language of rights and empowerment to women entrepreneurs. Feminists founded businesses along movement principles, such as publishing ventures that would give voice to women's words and perspectives.

During the 1980s and 1990s, women who wanted to start a business or help it to grow got a major boost from the explosion in the number of new opportunities to network with other successful women entrepreneurs and to learn about access to federal programs and resources. Women broke other barriers, launching firms in male-dominated industries such as finance, insurance, engineering and increasingly "green businesses". (From Ideas to Independence: A Century of Entrepreneurial Women, n.d.)

Technological innovation ramped up fast, as the 1990s became the 2000s. That not only enabled women entrepreneurs to break into technology-based businesses in record numbers, but also to use technology to start, run, promote and accelerate all types of companies.

Women are often typecast as better manipulators and appeasers than men, and are famed for their higher ability to see things from the perspective of others. (Harari,

2015). But in the real world, these stereotypes do not hold true. Over years of evolutionary processes, the social norms pushed women to be more and more submissive, while the men started becoming more and more aggressive. It is due to these social constructs that gendered perspectives have been created in the society deciding gendered roles for each of the genders.

Since men obviously outnumber women at the workplace, there are more 'male-bonding' sessions from which women are excluded. (Krishnan, 2013) This creates greater divide between men and women at work and even in the entrepreneurial market. The set social constructs restrict men in from of women and women in the presence of men, as they feel uncomfortable in such situations.

WOMEN ENTREPRENEURSHIP IN INDIA

As a patriarchal society, women have always played a second fiddle when it comes to their economic contribution. Any income generated by women is considered to be secondary, and not primary. Generation of primary income is a normative role shouldered only by the males of the society. This clearly leads us in to a conditioned belief that work is also gendered. The activities of daily living (ADL) are therefore divided between who does what as males or females, and working or enterprising as a role certainly does not fall in favor of women.

Historically, entrepreneurship in India is closely associated with the country's culture, customs and traditions, entrepreneurship often evolved along with societal etiquette. (Rizvi & Gupta, 2009) However, at present, the success of the individual as an entrepreneur depends mostly on his or her imagination, vision, innovativeness and risk-taking ability and need not be entwined with age-old cultural and societal etiquette. Rise of entrepreneurial needs can be considered to be closely linked to social, cultural, religious and psychological variables. Current entrepreneurship is thus centred on creative thinking and the growth of new ideas. However, whether this is the case for women entrepreneurs in India has not been fully explored. (Reddy, 1991) (Nayyar, Sharma, Kishtawaria, Rana, & Vyas, 2007)

At this juncture, it is worth mentioning that entrepreneurs are different from small-business owners who are primarily concerned with securing an income to meet their direct needs, rather than engaging in innovation. On the other hand, entrepreneurs display higher achievement motivation and greater risk-taking ability and typically provide employment to others. In view of these reports and statements, the lingering question is what issues are confronted by Indian women entrepreneurs, where entrepreneurship is believed to be a male prerogative. (Mathew & Panchanatham, 2011)

It is believed that entrepreneurship is ideal for women who are seeking to contribute their share of participation in the country's economic development. Since women are an important part of the industrialized society, it is now considered that women have to work using her acquired skill to supplement income for the family. Now they are likely to pick up self-employment opportunities rather than waiting for employment in the organized sector. For this, there is a need to create suitable alertness and environment to promote entrepreneurship.

Women entrepreneurship is gaining significance in India in the wake of economic liberalization and globalization. (Kumari, 2012) The policy and institutional framework for evolving entrepreneurial skills, providing vocation education and preparation has widened the horizon for economic empowerment of women. However, women account for only one third of the economic enterprises. There exists a surplus of successful business women entrepreneurs both in social and economic fields in India and they are performing significantly well. Government of India has also led National Skill Development Policy and National Skill Development Mission in 2009 in order to deliver skill training, vocational education and entrepreneurship development to the developing work force. The growing presence of women in the business field as entrepreneurs has transformed the demographic characteristics of business and economic growth of the country. Women-owned businesses enterprises are playing a more active role in society and the economy, inspiring academics to focus on this remarkable phenomenon.

The woman entrepreneur in India has various qualities. A woman or a group of women manage the whole business of enterprise. She formulates numerous plans and executes them under her own supervision and control. There may be some persons to help her but final control lies with the woman. A woman entrepreneur must offer at least 51 percent of the employment generated in her enterprise to women. A woman entrepreneur takes calculated risk. She faces uncertainty confidently and assumes risk. She has to tie up capital and wait for good returns. A woman entrepreneur likes to take accurate risks because she wants to be a successful entrepreneur. The most critical skill essential for industrial development is the capability of building a sound organization. A woman entrepreneur assembles, co-ordinates, organizes and manages the other factors namely land, labour and capital. It is essential to be a self-confident for a woman entrepreneur. She should have faith in herself and in her capabilities. She should have the confidence to implement the amendment and overcome any opposition to change. A woman entrepreneur should have courage to own the mistakes and correct them. The key function of a woman entrepreneur is to make decision. She takes several decisions regarding the activities of her enterprise. She decides about the kind of business to be done and the way of doing it. A woman entrepreneur must be clear and creative in decision-making process. A woman entrepreneur is one who incubates new ideas, starts her enterprise with these ideas and offers added value to society based on their independent initiative.

A unique feature of a woman entrepreneur is the willingness to work hard. She has to follow the standard; a woman entrepreneur is an achievement-oriented woman, not money hungry. She works for challenge, achievement and service to others. Achievement orientation is a derivation to overcome challenges, to

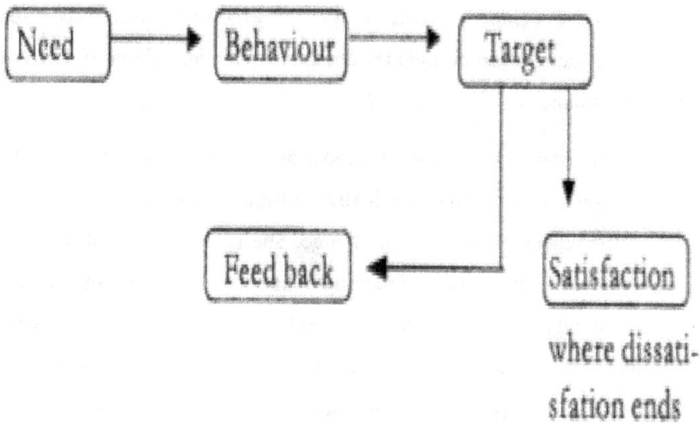

advance and to grow. A woman entrepreneur must be optimistic. She should approach her undertaking with a hope of success and attitude for success rather than with a fear of failure. The positive thinking of woman entrepreneur can turn the condition favourable to her.

The success of an enterprise mainly depends upon the ability of woman entrepreneur to cope with latest technology. (Mishra S. , 2009) Technical competency refers to the ability to devise and use the better ways of producing and marketing goods and services. Women entrepreneurs face the difficulties boldly and bravery. She has faith in herself and tries to solve the problems even under great pressure. A woman entrepreneur is active, focused, having a mission and a clear vision. She should be a woman of creative and analytical thinking. She must be intelligent, flexible and problem solver. Leadership quality is one of the most significant characteristic of a woman entrepreneur. It is the process of impelling and supporting others to work actively towards achieving objectives. (Brush, 2009)

There is no refuting of the fact that Indian women have come a long way, but according to facts exposed by the Sixth Economic Census by the National Sample Survey Organisation (NSSO), they still have a long way to go when it comes to matching the numbers with their male counterparts. (Sharma K. , 2016) According to the data collected during the survey, only 14 percent of business establishments in the country are being run by female entrepreneurs. This means, out of the 58.5 million functional businesses, only 8.05 million of them have a female as a chief. The data composed by the survey also discovered that most of these women run companies are small-scale and about 79 percent of them are self-financed.

When women comprise of almost 50 percent of the total population it is but natural that they will be gradually contributing their bit to the economy. A woman entrepreneur is an agent of change. She is an initiator of an enterprise. She is involved in the process of making links between opportunities and resources. She constantly strives for responding to the business environment with commitment and motivation. Entrepreneurship is a function of creating something new, organizing, coordinating, and undertaking risks and handling uncertainties.

Women entrepreneurs represent a group of women who have broken away from the beaten track and are exploring newer avenues of economic participation. Women with high drive, creativity, innovation and those who are keen on achieving on their own have taken up the challenging role of entrepreneurship. Women no longer confine themselves to conventional fields like embroidery, knitting and tailoring but are now venturing into new fields like electronics, pharmaceuticals engineering and services. These women are mostly urban, appropriately educated, sometimes having specialised skills and equipping themselves with some training and financial support. In spite of women taking entrepreneurship in many challenging fields, the current entrepreneurial activity in India is not very high. There exist a definite gap between the current status of women entrepreneurial activity and their potential. The social origins of women entrepreneurs play an important role in defining the kind of work they undertake. Women also find micro enterprises attractive because of low barriers to entry and the flexible nature of work, which makes it easy to combine gainful employment with domestic responsibilities. (Carr, 1990)

Entrepreneurial venture for women can be planned and developed. Women entrepreneurship paves the way for better and fuller utilization of capital and also mobilizes the potential female resources.

WOMEN ENTREPRENEURS IN GUJARAT

Business runs in the blood of every Gujarati. It has spread its acumen over USA, Canada and many other Western horizons. With this premise, the Centre for Entrepreneurship Development (CED) was established in 1970 in Gujarat to enhance trader's expertise and help the businesses flourish. In order to encourage more women entrepreneurs from the state, CED started various programmes to train women in this domain.

Ease of Doing Business Ranking 2019

Rank	Economy	EODB Score	EODB Score Change
1	New Zealand	86.59	0.00
2	Singapore	85.24	+0.27
3	Denmark	84.64	+0.59
4	Hong Kong SAR, China	84.22	+0.04
5	Korea, Rep.	84.14	-0.01
6	Georgia	83.28	+0.48
7	Norway	82.95	+0.25
8	United States	82.75	-0.01
9	United Kingdom	82.65	+0.33
10	Macedonia, FYR	81.55	+.032
13	Taiwan, China	80.90	+0.24
22	Canada	79.26	+0.38
24	Germany	78.90	0.00
32	France	77.29	+0.99
39	Japan	75.65	+0.05
77	**India**	**67.23**	**+6.63**

Source: PHD Research Bureau, Compiled from World's Bank Doing Business database

The above figure also shows that India ranks 77th on the Ease of Doing Business Ranking 2019.

The Government of Gujarat has taken many steps to encourage the women of the state to take up their entrepreneurial careers. While he was the Chief Minister of the state, Shri Narendra Modi ji kept on emphasizing in his speeches that entrepreneurship is inbuilt in the people of Gujarat. He further kept addressing that women play an important role for the economic development. According to him, even in the dairy sphere, the contribution of men is minimal. He suggested that women should be made an integral part in the decision making process in almost all the spheres.

According to the All India report of Sixth Economic Census 2016, there is presence of almost 6.57% entrepreneurial establishments set-up by women in Gujarat.

Category of Workers (Main & Marginal)

(i) Cultivators	Persons	5,447,500	21.99
	Males	4,244,449	23.58
	Females	1,203,051	17.78
(ii)Agricultural Labourers	Persons	6,839,415	27.61
	Males	3,649,591	20.27
	Females	3,189,824	47.14
(iii)Workers in household industry	Persons	343,999	1.39
	Males	210,561	1.17
	Females	133,438	1.97
(iv) Other Workers	Persons	12,136,833	49
	Males	9,896,313	54.98
	Females	2,240,520	33.11

Figure showing Census data of workers in Gujarat.

From the above figure, it is evident that almost 33% females fall in the category of other workers. All persons engaged in 'work' defined as participation in any economically productive activity with or without compensation, wages or profit. (Menon, 2015). The Main workers are classified on the basis of industrial category into the following four categories:

1. Cultivators
2. Agricultural Labourers
3. Household Industry Workers
4. Other workers

Thus, there is a substantial evidence of women engaged in entrepreneurship within Gujarat. Engagement in entrepreneurship can mean either they are economically engaged in their family enterprise, are partners in business or they run an enterprise on their own. The focus of this study is only those women who are successfully running enterprises on their own within the state of Gujarat. The Gujarat Chamber of Commerce and Industries (GCCI) has a special Business Women's Wing (BWW), which has enlisted women entrepreneurs within

Ahmedabad. Similarly, The Southern Gujarat Chamber of Commerce and Industries (SGCCI) also has a Ladies Wing that has listing of Women Entrepreneurs within Surat City. The Vadodara Chamber of Commerce and Industries (VCCI) is working on collating data on women entrepreneurs of Vadodara whereas the Rajkot Chamber of Commerce and Industries (RCCI) does not have any such provision.

The number of women-owned businesses have grown tremendously and are proving to be a successful segment of the business world. (Mehta & Parekh, Women Entrepreneurs: Leading the Way Ahead, 2014) According to the Women's Global Entrepreneurship Study conducted in USA, UK and India, commissioned by Dell (2012), it has been found that the ideal country for women starting a business in 2012 could well be India. (Lal, 2014)

Women are to be reached and motivated, so that they extend a helping hand to men in their entrepreneurial venture and the women too, take some enterprises. The entrepreneurial women are simply the women who want to start their own business. The most significant characteristic associated with entrepreneurs are the need for achievement, support, independence and leadership. Hence, these aspects need to be fostered and developed among women.

PURPOSE OF THE STUDY

The idea of women entrepreneurship is becoming a worldwide phenomenon. Women access to knowledge, skills, resources, opportunities and power but still remain rather low especially in the rural areas. Although women remain involved in family trade but they neither are in organized form nor do women have its ownership making their role as a mere helper. Women of today are clearer in their life-goals. They know how to undergo all the hardships as well as how to cherish the success which comes after. In our culture, women are socially conditioned to be submissive and not speak much. But a woman with entrepreneurial ambitions must make herself heard. While making tall claims on achieving women empowerment, we must also provide opportunities and the right kind of platforms to women to achieve. Women have been taking growing interest in recent years in income producing activities, self-employment and entrepreneurship. This is seen in respect of all types of women both in urban and rural areas. Women are embracing both traditional activities and also non-traditional activities. It is clear that progressively more women are coming forward to set up enterprises. In the course of entrepreneurship, women have to face many problems associated with entrepreneurship and these problems get doubled because of her double role as a wage earner and a homemaker. It becomes all the more significant to study what keeps these women well motivated in order to attain their set goals and targets. It is also significant to study what leadership forms are followed by these women in order to handle their human resources and other resources as well as how they balance their work-life while setting up their enterprises.

The profession of social work has always emphasized on empowering women and it is considered to be one of the most important thrust areas of the profession. Encouraging livelihoods and promoting skill-development has always been of prime importance to the field of social work. Every trained social worker aims to uplift the status of their stakeholders in every possible way. Promoting gender equality and encouraging gender-neutral workplaces and work environment is the primary principle of social work practice. With the dawn of the new century, women have sought to look more beyond just a stable job. The idea of women entrepreneurship is becoming a global phenomenon. Women access to knowledge, skills, resources, opportunities and power but still remain rather low especially in the rural areas. Although women remain involved in family trade but they are neither in organized form nor do women have its ownership making their role as a mere helper. Women of today are clearer in their life-goals. They know how to undergo all the hardships as well as how to cherish the success which comes after. In our culture, women are socially conditioned to be submissive and not speak much. But a woman with entrepreneurial aspirations must make herself heard. While making tall claims on achieving women empowerment, we must also provide opportunities and the right kind of platforms to women to achieve.

An entrepreneur plays a very dynamic role for his enterprise. A woman, considered vulnerable, has to combat a lot of hurdles and obstacles which come in her way achieving her goals. The researcher will also consider the environmental challenges, strengths and weaknesses of women entrepreneurs and that how under even stressful times, a woman exhibits strength in holding her career and personal life together.

Women entrepreneurs have to play a dual role in their lives. They are supposed to undertake household chores as well as manage the enterprise. If a woman seems to be too ambitious towards her career, she is labelled as insensitive towards her homely responsibilities. Women forget that they too have a capacity in trying to fulfil their dual roles. This feeling of being a super-woman, gives way to conflicts. Career women often have to face work-home role conflict and so the concept of work-life balance becomes an important parameter to be studied while researching on women entrepreneurs.

Working women and especially enterprising women need to be authoritative and showing some leadership skills. It also becomes necessary to study the leadership aspect of enterprising women as that factor helps them retain manpower and expand their enterprise. The researcher will also study the leadership styles adopted by enterprising women and how it affects their entrepreneurial performance. Entrepreneurs are the real heroes of economic life. They are the ones who sweat and toil and sacrifice and that natural resource gain value only by the ingenuity and labour of man. From the knowledge of failure, they forge success. In accepting risk, they achieve security for all. In embracing change, they safeguard social and economic steadiness.

Motivation

All the human beings have a desire to have one or more needs to be satisfied for one or another reason. This happens due to the result of dissatisfaction of present situation. In other words, need arises from the dissatisfaction. The person is focused to do something to satisfy the need, and therefore he put his all the efforts till the need is satisfied fully. During this period, through feedback he changes his behaviour so that he can work in right direction to attain goal. As soon as the goal is achieved the dissatisfaction is over, but the needs are constant and infinite, therefore the process remains continued.

Motivation is a behavioural related general phenomenon, but to motivate entrepreneur there are some social goals which are listed here under:

1. Achievement
2. Power
3. Affiliation
4. Independence
5. Extension
6. Personal achievement
7. Social achievement, etc.

From the above listed goals for entrepreneurship development, the following three goals are very important:

1. **Need for Achievement (n-Ach):** A strong desire of an individual to attain the standard of excellence.
2. **Need for Power (n-Pow):** Some people are crazy for power. One can motivate them by giving them more and more power or authority. Such persons can be a manager or supervisor.
3. **Need for Affiliation (n-Aff):** Some people desire to create, increase and maintain the relations with others. Such people are generally devoted and always try to avoid the conflicts.

It is important to note that David McClelland has carried out several studies in India, Malawi and Ecuador in this line. He has come to the conclusion that for the development of the entrepreneurship, achievement motivation is very important factor.

In a study on motivating factors among Indian Entrepreneurs, the factors prompted them to promote their companies were divided into two major sub-heads viz., internal factors and external factors.

Motivating Factors

Internal

* Educational Qualifications
* Occupational Experience
* Desire to work independently in manufacturing line
* Desire to branch out to manufacturing from present occupation.
* Family background
* To earn profits and to possesses wealth
* To engage family members along with himself
* To possess social prestige

External

* Assistance from Govt.
* Assistance from financial Institutions
* Availability of technology/ raw materials
* Demand of a particular product
* Wanted to utilise excess money
* financial help from non-government sources.

(Figure showing Motivating Factors of Entrepreneurs)

The main conclusion of this research was that the behaviour of a matured person can be directed towards the best results by developing his need for achievement of goal. Of course, the conclusion of the studies and researchers carried out afterwards have stated that to motivate the entrepreneurs following bases regarding the behaviour are also very important.

1. Tolerance to ambiguity
2. Problem solving
3. Creativity, etc.

Achievement Motivation

Entrepreneurship is a major factor in the national economy; thus, it is important to understand the motivational characteristics spurring people to become entrepreneurs and why some are more successful than others. (Collins, Hanges, & Locke, The Relationship of Achievement Motivation to Entrepreneurial Behavior: A Meta-Analysis, 2004)

The concept of need for achievement (n-Ach) was formulated in the 1950s. McClelland and his colleagues argued that high n-Ach people are more likely than low n-Ach people to engage in energetic and innovative activities that require planning for the future and entail an individual's responsibility for task outcomes. The concept of achievement motivation has been used to describe individual efforts towards personal goals in social surroundings (Cassidy & Lynn, 1989). It is also found to be a typical characteristic among individuals who are successful in their work (McClelland, The Achieving Society, 1961) (McClelland, Money as motivator: some research insights., 1967). A strong achievement motivation is probably the one characteristic most clearly describing the attitudes and the behaviour of an entrepreneur (Virtanen, 1997). It may also be considered as a dimension closely related to other typical characteristics for an entrepreneur, such as willingness to take risks (e.g. Littunen 2000) and growth orientation (Kirschenhofer, 2006). McClelland (1967) suggested that the need for achievement is one of the basic characteristics of an entrepreneur, which may be typical for certain cultures as well as learned. Several researchers argue, however, that McClelland's conception of the achievement motivation only covers one aspect of the myriad of dimensions affecting it. Cassidy and Lynn (1989) made themselves conversant with the measurements used in various studies, and recognised seven different dimensions affecting an individual's achievement motivation. According to them, an individual's achievement motivation consists of work ethic, acquisitiveness, dominance, excellence, competitiveness, status aspiration, and mastery.

While studying entrepreneurial motivation, Murthy et. al., studied and classified the motivating factors on different basis. According to them, entrepreneurs are motivated to start business enterprises due to the following three types of factors:
(1) Ambitious factors
(2) Compelling factors
(3) Facilitating factors.

McClelland says that in a portable society where an occupational position is somewhat dependent upon performance (rather than on family or political connections) managers/executives should have higher need for accomplishment, than men in other occupations.

As far as **Achievement Motivation** is concerned, there is a great sense of achievement in women, and on knowing that people like them have achieved much in life, women are also motivated to work harder. This also hints at the aspiring and hardworking nature of the women respondents.

It becomes imperative to study the Achievement-Motivation of women in order to understand the reasons that motivate women to be enterprising despite the social stigma associated to it being an area suitable for men.

Leadership

Leadership is a procedure whereby an individual influences a group of individuals to accomplish a common goal. The components consist of a process involving influence, which occurs within a group context focusing on an ultimate goal. Leadership style is characterized by the consistent patterns of exhibited behaviours, attitudes, and values. Environmental or situational forces that are continuously changing or evolving affect these patterns. Some classification style examples include participative, consultative, consensus, democratic, autocratic, entrepreneurial, and contingency leadership (Dubrin, 2004).

Leadership has been defined as traits, behaviours, influence, persuasion, interaction patterns, role relationships, and as administrative position; and most definitions assume that leadership involves a process whereby intentional influence is exerted by one to others in an attempt to guide relationships (Moerer-Urdahl, 2005).

Albert Einstein rightly quoted, "The woman who follows the crowd will usually go no further than the crowd. The woman who walks alone is probable to find herself in places no one has ever been before." In order to tread on uncharted paths alone, a woman has to endure a lot of struggles and hardships before becoming successful. During times of struggles and trials, it is only strong leadership skills that help women to stay grounded and focussed.

Leadership is an important virtue for entrepreneurs as it will help the entrepreneurs to hold their workforce together during every stage of the development of business. A good leader will be able to bring along all the people along with him/ her whereas a leader who cannot lead his/ her people well will gradually end up losing resources. For an entrepreneur, resources are of prime concern. It becomes the duty of an entrepreneur to also be good leaders who believe in the potential of the people who work with them.

Women surpass men in leadership effectiveness when using the transformational leadership behaviours (Fondas, 1997). The differences between styles of leading within the gender arena are significant to women's progress, or lack thereof, in business as women leaders continue to hit their heads on the "invisible" glass ceiling. "The glass ceiling is a global phenomenon whereby women are disproportionately concentrated in lower-level and lower-authority leadership positions than men. Gender is essential to contemporary notions of effective styles that have transformed from a traditional masculine, autocratic style to a more feminine and androgynous style of democratic and transformational leadership (Knopik & Moerer, Women Leaders and Entrepreneurs: Learning from One to Teach the Other , 2008).

Work Life Balance

Work-Life Balance does not mean an equal balance. It means the ability to schedule the hours of professional and personal life so as to lead a healthy and peaceful life. It is not a new concept. It highlights the values, attitudes and beliefs of women concerning their age to work in organizing and balancing their work and personal life. (Deshmukh & Deshmukh, 2016)

In the last two decades Indian women have entered work force in large numbers and many of them hold senior positions now. Women are also increasingly starting businesses of their own. Studies have shown that the majority of women work for approximately 40-45 hrs./week out of which almost 53% of them struggle to achieve work-life-balance. The reason behind this struggle is that they are being challenged by the demands of their organization versus the commitments of their home. They need to manage the daily requirements of their family as one side and the multiple schedules, meetings, business requirements and other routine responsibilities at work. One major challenge for enterprising women is the difficulty of combining work with a satisfactory personal life. Moreover, it has been traditionally believed that a good career makes a good home impossible and so the task becomes ambiguous and ambitious. Women turn to entrepreneurship for various reasons especially for satisfying their personal needs of self-expression an autonomy. This also makes their lives increasingly complex with pressure, intense working hours and lesser involvements in the family hence they juggle to satisfy both the roles. (Rani, 1996) They may receive additional help at home in the form of servants, parents, in-laws and husband but still the entire responsibility of the home rests with the woman. But the problems remain, as the entrepreneur is not a super woman and there follows a conflict.

Women have always been burdened with the responsibility of the domestic chores. The primitive society had mandated women to stay at home and look after the home affairs while the men would go out and work in order to financially support their families. While the process of evolution was on, feminine genes which were passed on to the future generations started becoming more and more submissive whereas the masculine genes became more and more aggressive as they were conditioned to be such.

So, in today's times also, it becomes beyond comprehension of the society when women step out of the realms of the social norms in order to assert themselves and compete with their male counterparts. However, being conditioned to work for home; women also feel a sense of guilt when they pay any less attention towards their homely duties. Hence, no matter what profession they choose, they have to prioritize their homely duties in order to match it with their professional duties.

For women in an organizational set-up where the work timings are fixed, it becomes easier to divide their time in order to fulfil both the requirements; however, for enterprising women, it becomes a bit difficult especially while they are setting up their venture and/ or trying to meet their deadlines. They need to focus on their enterprise more than anything else, if they intend to survive the cutthroat competition and succeed.

Hence, this aspect of work-life balance of women entrepreneurs also becomes important to study so that it can be known how they manage all their domestic chores and yet shine in the entrepreneurial world.

Relevance of this study to Social Work Profession

According to International Federation of Social Workers (IFSW), Social Work is a practice-based profession and an academic discipline that promotes social change and development, social cohesion and the empowerment and liberation of people. Principles of Social Justice, Human Rights, Collective Responsibility and Respect for Diversities are crucial to Social Work. It simply means that Social Work is a profession which is inclusive in its nature and approach. Its core values hold inclusive development and acceptance of people.

Entrepreneurship is a field which has witnessed many successful women. From self-help groups and micro financing to single-handedly managing the entire business, women have achieved it all. And this is no small feat for women who come from a culture which has often and always pulled them down and mocked upon their social status.

Women have always withstood the worst of social changes and therefore labelled as a vulnerable section of the society under social work. Lot of studies have been undertaken in order to chart out ways to empower and uplift women. An entire field of study - Women's Studies, is propagated to understand women's perspectives to development. Under the Women's Studies, lot of empowerment and development has already been done for the women who were considered as an excluded lot and a burden to the society.

We live in a cultural array, which is heavily dichotomous in nature. On one hand, we have Goddess *Laxmi* as a deity of wealth and on the other hand, the same *Laxmi* is considered as a burden to the family in which she is born due to the prevalence of dowry system. We worship Goddess *Saraswati* as the Goddess of Intellect; yet, we fail to create equal opportunities for human *Saraswati*s when it comes to pursuing education. We take pride in the triumphant feat of Goddess *Durga* over the demon *Mahishasura*, yet we fail to acknowledge the tremendous feats achieved by all the *Durga*s around us in their struggle to assert themselves in this heavily patriarchal society.

Despite these social oddities, determined women have carved their ways out towards development. From struggling for their basic right to live and participate to successfully running empires, women have achieved it all. Many a times, like-minded people of the opposing genders did help women to succeed, but many a times, these successes were purely on the basis of sheer hard work, determination, focus and belief in self.

However, in this fast-paced world, the challenges are of a different kind. Women may have reached to the pinnacle of success but not without sacrificing their family life or other core values. It is with this intention that this study needs to take place so that the expectations, ideas and aspirations of the enterprising women of the millennium can be understood and appropriate assistance can be provided to foster their skill, capacity and talent.

This sets the tone for studying the various psychosocial aspects concerning women entrepreneurship. Hence, the researcher has chosen to study three factors - Achievement Motivation, Leadership and Work life Balance, out of the many factors under the purview of this study.

RESEARCH SETTING

In the state of Gujarat, "entrepreneurship" is revered almost as a religion. This is the land where entrepreneurs have made their own fortunes with sheer hard work and dedication. The inherent spirit of adventure and the huge appetite for risk have led the *Gujaratis* to phenomenal success in their businesses. The entire ecosystem of Gujarati culture works around entrepreneurship. Entrepreneurship runs in the blood of a Gujarati. Children in a Gujarati family are exposed to money making businesses early on. Even during social gatherings people talk about business rather than bureaucracy, politics, or literature. By the time a person comes out of college he would have a role model in one or other successful businessman. For a potential entrepreneur in Gujarat, the choices of role-models are also not less. Everyone wants to become Adani, Ambani, Khambatta, Kotak, Patel, Shanghvi, and the list goes on. With this premise, the researcher has based the study within the state of Gujarat with an emphasis to understand the entrepreneurial activities of the women of the state. The researcher narrowed down to focusing on the four major cities of Gujarat – Ahmedabad, Rajkot, Surat and Vadodara for greater availability of respondents.

Ahmedabad was once known as the Manchester of the East owing to the presence of various cotton mills. In 2010, Ahmedabad was ranked third in *Forbes*'s list of fastest growing cities of the decade (Kotkin, 2010). This city also houses various entrepreneurial ventures especially run by women that have gained national exposure. The famous Jasuben Pizza, which was popularized by Shri Narendra Modi ji is also from Ahmedabad. The Gujarat Chamber of Commerce and Industries (GCCI) has a Business Women Wing (BWW) which has membership from business women and women entrepreneurs of the city. Hence, the researcher chose Ahmedabad as one of the cities for data collection.

Rajkot is the fourth-largest city in the state of Gujarat. It is the second-cleanest city of India, and according to a CityMayor Statistics report published in 2018, Rajkot is the 22nd-fastest-growing city in the world as of July 2019. As a city effervescent with growth and empowering women, the researcher chose this city as a study area.

The third city chosen for this study Surat, popularly known as the diamond capital of the state. It was once known to be the largest seaport famous for trading. Surat is famous for its food all over India. There is a famous saying in Gujarati language "સુરતનું જમણ અને કાશીનું મરણ", meaning *Eat in Surat and Die in Kashi* for the ultimate experience and repose to the soul. Surat houses many industrialists and diamantaires. As much as the men are involved in businesses, Surati women are also not behind. The Souther Gujarat Chamber of Commerce and Industries has a dedicated Ladies' Wing to facilitate the business women and women entrepreneurs. The researcher was greatly facilitated by the presence of this agency and hence the researcher chose to undertake Surat as one of the study settings.

Vadodara is the researcher's hometown and has multiple possibilities for women entrepreneurship. Baroda State was a former Indian Princely State ruled by the visionary Gaekwads. Maharani Chimnabai, wife of His Highness Sayajirao Gaekwad III, was one of the pioneers of establishing self-help groups for women in the city of Baroda much before it was conceptualised in the rest of the world. She worked in the areas of female education and women empowerment. Thus, the researcher has also considered Vadodara as one of the study settings.

Overall, each of the four cities of Gujarat have contributed to the idea of women empowerment through entrepreneurship. Therefore, it becomes important to cover these four cities in order to understand the change in trends to women entrepreneurship within the state of Gujarat.

CHAPTER II
REVIEW OF LITERATURE

REVIEW OF LITERATURE

In the earlier chapter, the researcher has discussed about the various aspects of entrepreneurship, particularly female entrepreneurship and why it is the focus of study.

In this chapter, the researcher has provided various studies which have been undertaken at International and National levels with specific focus to fostering women entrepreneurship. The researcher has also gathered various published articles on women entrepreneurship and incorporated them in this chapter.

On a concluding note, the researcher has provided the significance of her research based on the research gaps found in all previous studies.

Concept of Entrepreneur

The term entrepreneur is being defined in many ways. The concept of entrepreneur varies depending on the level of development, thoughts and perception as well as from one time period to the other. Initial effort to define entrepreneurship is found in the writings of Richard Cantillion (1755) who defines entrepreneur as a person who pays a certain price for a product to resell it at an uncertain price, thereby making decisions about obtaining and using resources while consequently assuming the risk of enterprise. A significant face in Cantillion's argument is that entrepreneurs would always look for the best opportunities for using resources for their highest commercial yield leading to economic growth and development. Accordingly, if an individual can grab the opportunity, then s/he has trait of becoming entrepreneur. Similar definition is found in Adam Smith's writings (1776), whereby he defines an entrepreneur as an individual who undertakes the formation of an organisation for commercial purpose. According to him, an entrepreneur is a person with unusual foresight who could recognise potential demand for goods and services. According to him, entrepreneurs respond to the economic change, thus becoming the economic agents who are responsible for transforming demand into supply taking care of the needs of the economy and society.

Say (1803), described entrepreneur as one who possessed certain arts and skills of creating new economic enterprises, yet a person who had exceptional insight into society's needs and was able to fulfil them. In this definition, Say has combined the 'economic risk taker' of Cantillion and 'industrial manager' of Smith into unusual character. Say's entrepreneur is influenced by creating new enterprises and at the same time is inclined by society's requirements and fulfils them through rational management of resources. According to him, entrepreneur tends to have both the qualities rather than just one quality.

Mill (1848), described entrepreneur as a business founder, and this 'fourth factor' of economic endeavour has been ingrained in economic literature as encircling the final ownership of commercial enterprise. Therefore, Mill's entrepreneur is the owner and founder of his/her business enterprises. This also means that in absence of this fourth factor of production an economy cannot develop and prosper.

Menger (1871) also defines entrepreneur is a change agent who transforms resources into useful goods and services, often creating the circumstances that lead to industrial growth. Menger also looked upon entrepreneur as a smart individual who could visualize this transformation and generate the means to execute it. "Entrepreneur is able to see both extremes and envision ways to transform the raw material into a useful commodity". According to Menger, entrepreneur should have vision to foresee the future, transform, and produce useful commodity.

Schumpeter's definition **(1961)** saw a turn in the concept of an entrepreneur whereby he defined an entrepreneur as a dynamic agent of change; or the catalyst who transformed increasingly physical, natural and human resources into corresponding production possibilities. According to Schumpeter, "entrepreneurship connotes innovativeness and an urge to take risk in the face of uncertainties and intuition, i.e., a capacity of seeing things in a way, which later prove to be true".

All the above-mentioned definitions prove that there has been considerable research on what promoted entrepreneurship in different societies and economies. Different definition at different point of time emphasize on a variety of factors under different circumstances. In addition, these definitions show that various economists emphasized the existence of economic opportunities and incentives as essential for progress of entrepreneurship. The critical analysis of these various definitions however, brings out certain common characteristics. These include the perceptions of economic opportunities, technical and organisational skills, managerial competence, and motivation to achieve results. The entrepreneur has a potential to envision future and exploring the opportunities for gainful purpose. In addition, the various concepts and theories propounded by researchers appear to point that the emergence of entrepreneurs in any society directly depends upon interlinked economic, social, cultural and psychological variables.

Women Entrepreneurship

The definition of entrepreneurship has never been differentiated on the basis of gender and can be extended to women entrepreneurs without any restrictions. "Women entrepreneurs are defined as the women or group of women who initiate, organize and ran a business enterprise. In terms of Schumpeterian concept of innovative entrepreneurs, women who innovate, duplicate or adopt a business activity are called women entrepreneurs" (Khanka, 2006). So women also have the same abilities like men to start a business venture and run their enterprise successfully.

Government of India defines a woman's enterprise as "an enterprise owned and controlled by a woman and having a minimum financial share of 51 percent of the capital and giving at least 51 percent of employment to women". However, this definition may prove to be practically dysfunctional. The simple reason is, women entrepreneur may like to employ male workers sometimes or it may so happen that she is managing business at her own without employing any one in her enterprise.

Similarly, **Ganesan** also defines women entrepreneurs as women or a group of women who initiate, organize and operate a business enterprise. "Women who innovate, initiate or adopt an economic activity can be called as women entrepreneurs" (Ganesan & Duraipandian, 1999)

The above-mentioned definitions consider that women are also innovative, can organise and operate business. They are in no way different from their male counterparts as far as managing business is concerned. Thus, a woman can also become a successful entrepreneur who is capable of managing her enterprise, and thereby contributing significantly to the economic health of the country.

Women Entrepreneurship has been a recent concern. The number of women-owned businesses have grown tremendously and are proving to be a successful segment of the business world. (Mehta & Parekh, Women Entrepreneurs: Leading the Way Ahead, 2014) According to the Women's Global Entrepreneurship Study conducted in USA, UK and India, commissioned by Dell (2012), it has been found that the ideal country for women starting a business in 2012 could well be India. (Lal, 2014)

Women are to be reached and motivated, so that they extend a helping hand to men in their entrepreneurial venture and the women too, take some enterprises. The entrepreneurial women are simply the women who want to start their own business. The most significant characteristic associated with entrepreneurs are the need for achievement, support, independence and leadership.

Characters in social status of women are inevitable. Psychological dependency of the business women on their family members in decision making. To share the family responsibility along with entrepreneurial responsibility. The personality traits of the women in communicating with others. Lack of interest and proper exposure to the things leading to run enterprise with binomial names of women. Lack of proper training before entering into the business. Interested in routine matters only and not involving in innovative ventures. Lack of marketing orientation in entrepreneurship. Inability to differentiate entrepreneurial functions from other functions like management, production and speculation.

Women are key drivers of the Indian start-ups.

Category of Workers (Main & Marginal)			
(i) Cultivators	Persons	5,447,500	21.99
	Males	4,244,449	23.58
	Females	1,203,051	17.78
(ii)Agricultural Labourers	Persons	6,839,415	27.61
	Males	3,649,591	20.27
	Females	3,189,824	47.14
(iii)Workers in household industry	Persons	343,999	1.39
	Males	210,561	1.17
	Females	133,438	1.97
(iv) Other Workers	Persons	12,136,833	49
	Males	9,896,313	54.98
	Females	2,240,520	33.11

(Source: PCA 2011)

Figure showing Census data of workers in Gujarat.

From the above figure, it is evident that almost 33% females fall in the category of other workers. All persons engaged in 'work' defined as participation in any economically productive activity with or without compensation, wages or profit. (Menon, 2015). The Main workers are classified on the basis of industrial category into the following four categories:

1. Cultivators
2. Agricultural Labourers
3. Household Industry Workers
4. Other workers

Thus, there is a substantial evidence of women engaged in entrepreneurship within Gujarat. Engagement in entrepreneurship can mean either they are economically engaged in their family enterprise, are partners in business or they run an enterprise on their own. The focus of this study is only those women who are successfully running enterprises on their own within the state of Gujarat. The Gujarat Chamber of Commerce and Industries (GCCI) has a special Business Women's Wing (BWW), which has enlisted women entrepreneurs within

Ahmedabad. Similarly, The South Gujarat Chamber of Commerce and Industries (SGCCI) also has a Ladies Wing that has listing of Women Entrepreneurs within Surat City. The Vadodara Chamber of Commerce and Industries (VCCI) is working on collating data on women entrepreneurs of Vadodara whereas the Rajkot Chamber of Commerce and Industries (RCCI) does not have any such provision.

In order to get relevant data on women entrepreneurs, the research scholar put in efforts on finding out the factors that motivate women to be enterprising, the socio-economic factors impacting the enterprises as women run them, the various leadership aspects that women use and the way they balance their work life and their personal/ family life.

Following are the various studies the research scholar has come across. These studies are categorised into two major categories - International Studies and National Studies. Within these two broad categories, the research scholar has incorporated material from various publication sources like books, journals, reports and data available online. The total number of International studies are 18; the total number of National studies are 54. The worker has reviewed 11 journal articles and working papers and some unpublished work has also been referred to. These studies are arranged chronologically to show how time improvised the research approaches to women entrepreneurship.

International Studies

Martin and Roberts (1984) states that women with children often have a more continuous work life, with spells of paid work between children and an earlier return to work after the youngest child reaches school age.

Naisbutt (1985) suggests that certain opportunities and products are suited for women. The thinking is based on the assumption that the distinctly female or male imagery associated with certain industries and business sectors continue to endure and influence who does what.

Hisrich (1988) in his work pertaining to impact of strategic origin of women on growth of their enterprises reveal that a woman entrepreneur who has previous experience in the field stands a better chance of successfully establishing her venture. Financial skills are of key importance to business survival.

Holmquist and Sundin (1988) observed that **female entrepreneurs seem to be more adjustable and adaptable than males.** Women working with husbands adapt to the business chosen by men. The most encouraging conclusion is that the females seem to be satisfied with entrepreneurship. The sense of independence and self-reliance is strong.

Kaplan (1988) in her work relating to women entrepreneurs observe that although their orientations and skills may vary, the large number of respondents are more concerned with creating employment for themselves rather than developing into entrepreneurs as defined by scholars. Non-economic reasons were often as important as the economic ones. Also, younger women found to be better educated, have better business skills and are more likely to plan and set goals; the impact of these factors on business sales, growth and profitability is less clear.

Nelson (1991) in his study on **'Small Business Opportunities for Women in Jamaica'** revealed that women were concentrated in businesses which required the least capital outlay or which an extension of household activities was. Majority of women had experienced sex bias while establishing and developing. Thirty person of women however experienced an advantage being a woman.

Harper (1992) in her work with poor communities of Orissa showed that the social and academic determinants that motivate women to cross borderlines from generating no income to generating small income through self-employment, and the borderline from there to growth-oriented micro or small business, found that one major barrier for women to star or expand business is the lack of exposure of business started and their approach towards it. They had pursued traditional career paths with only a minority in the non-traditional path.

Carter and Cannon (1992) in their study on **'Women Entrepreneurs'** in The Britain find out that the way women approach starting an enterprise is dominated by the stage they have reached in their life cycle, that is, their age and domestic relationships. Differentiation by personal life cycle is important as women start businesses at very different stages of life.

Hisrich and Fuldop (1993) in their work related to women entrepreneurs of Hungary found that most of the samples were married; the majority comes from middle or low income group. They were more educated than others were and were slightly more educated than their husbands were. They ventured due to their own interest in the area or due to job frustration.

Saleh (1995) observes that 75 percent of women entrepreneurs in his sample were degree holders; while in Sri Lanka, most successful women entrepreneurs belong to families with economic resources. Regarding the social background, a study of women entrepreneurs in Nepal indicated that women from the Tibeto-Burman communities are socially less constrained than those belonging to the Indo-Aryan communities in terms of entrepreneurial tendencies.

Junk Nancy (1998) in her article published in journal **'Work and Occupations'** of February vol. 25(1) examines the restrictive dimensions of self-employment. The analysis is based on in-depth interviews with 46 individuals, predominantly women, who work in home-based, self-employed professional, domestic services or production ventures.

Goheer (2002) in **'A survey of women entrepreneurs in the formal sector in Pakistan'** showed that most entrepreneurs were in the age group of 20-39 years. It also showed that the likelihood of a Pakistani woman being in business was greater if she lived in a nuclear family system, while the predominant mode in Pakistan is an extended family structure. Living in a nuclear family would mean that women have relatively less interaction with the older generation and are less constrained by the social and cultural influences. The literacy rate of women entrepreneurs and their family members were quite high; 97 percent of those were literate – mainly graduates and post-graduates. It shows that most women were engaged in traditional business while minorities were in non-traditional businesses, with textiles and apparel being the prime area of activity.

Carter, Sara and Cannon (2006) in their study on "Women Entrepreneurs-A Study of Female Business Owners: Their motivations, experience and strategies for success", found women to be more effective in downstream industries such as service and retail more frequently than men".

Verheul, Carree and Thurik (2007) in their study entitled **'The Normative Context for Women's Participation in Entrepreneurship: A Multicountry Study'**. This study was analyses the participation of women in entrepreneurship and examines the impact of specific norms of different countries that support women's entrepreneurship. The study chose 41 countries for analyzed. Global Entrepreneurship Monitor also found that the males' rate of entrepreneurial activity was three times more than that of females in some countries. So the participation of women in entrepreneurial activities can only be triggered by the initiatives taken by their country's respective governments as well as through an increase of the normative support for female entrepreneurship. The study concluded that a better understanding of the influencing cultural norms and values coupled with a change in the traditional beliefs embedded in the society are essential to increase the level of female entrepreneurial activity.

Global Entrepreneurship Monitor Report on Women and Entrepreneurship (2005) Women show nearly double the rate of early-stage entrepreneurship in middle-income countries than in high-income countries, with similar rates of business ownership activity regardless of country cluster. In addition, while women are most frequently motivated to become entrepreneurs by opportunity, necessity entrepreneurship is a larger motivator for women in middle-income countries. The gender gap is apparent across all countries and while it is somewhat narrower in middle-income countries, these countries are also places where the probability of a woman's business surviving is knowingly lower. Fear of failure plays a bigger role for women. Women's entrepreneurial ventures are most regularly found in the consumer-oriented sector. A shift toward the business services sector is evident in high-income countries. Women who are entrepreneurs have a tendency to fall in age between 25–44 years. It is likely that employment provides access to resources, social capital, and ideas that may aide in establishing an entrepreneurial venture. Education also is an important lever for accelerating such activity.

Countries on Role of Women in Development reports quotes, "The two facets of the role of women in production and reproduction should be conceived as complementary and interactive. In order to release women's potential for fuller participation in the productive and decision making processes, there is a need not only for a sharing of parental responsibilities but also for institutional provisions that would benefit children and families. Women should have the right, the requisite knowledge and the resources to regulate their reproductive capacity. However, without economic independence, women cannot achieve equality. Therefore measure to expand satisfactory employment opportunities to improve their economic status and provision of the opportunities, to improve their economic status and provision of the necessary infrastructure should be treated as an important and integral part of national and international development strategies."

Renana and Tate in their book **"Out of the shadows: Home based workers organize for international recognition"** suggests that home-based work in all of its diverse forms cannot be viewed as an artefact of traditional economics. Rather such work appears to be growing in modernizing rural economy. Further, the income it produces is not supplementary but rather increasingly vital to families and nations alike. The women who embroider are inextricably linked with the world's largest market.

Reheim S. in his article on 'Problems and prospects of self-employment as an Economic independence option for welfare recipients' says that social policy makers have begun to explore micro enterprise, as a route of welfare. With the passage of the Personal Responsibility and Work Opportunities Reconciliation Act of 1996, it is important for social workers to understand the benefits and the limits of entrepreneurs' growth so they can more efficiently influence welfare reform in the states. It also examines the problems associated with pursuing entrepreneurship as an option for achieving economic self-sufficiency and discusses the lessons learned from national studies of self-employment development programs. These studies suggest that entrepreneurship can be effective strategy for income generation and asset building for welfare recipients when the necessary supports are provided and policy barriers are removed. The potential impact of various approaches to welfare reform on self-employment is examined and recommendations for increasing the viability of self-employment as a route out of poverty are discussed.

National Studies

Lalitha D. (1982) in her book titled **'Status and Employment of Women in India'** tried to show that employment per se as against age, education, family type and place of residence plays a crucial role in raising the status of women. Employed women have more power and influence in the family and outside than unemployed women. The author made the analysis of the status within the family, at work place and in the wider society. There is hidden potential of employment that contributes to the status of women.

Singh and Gupta (1984) in their study on **'Potential Women Entrepreneurs'**, found out the reasons for becoming an entrepreneur amongst the respondents. 'Economic gain' ranked as the reason of greatest importance followed by 'Keeping oneself busy'. 'Fulfilment of one's ambition' ranked third followed by 'Wanted to become independent' and such reasons. In the motivational pattern, only 13 percent of women had a high score in need for achievement while 50 percent of them had moderate score and remaining had a low score. 58 percent of the women out of the total sample of high and moderate scores gave reasons as 'Wanted to become independent' or 'pursuing my own interest'.

Singh et al. (1985) in their work of **'Successful Women Entrepreneurs'** found out the reasons or motivational factors for venturing into entrepreneurship. To keep busy, to fulfil ambition and to pursue own interest were ranked the first three respectively, followed by 'by accident' and other factors like earning money, to provide employment etc. 70.8 percent had been taking male help in running the enterprise and only 29.2 percent were operating entirely on their own. 54.1 percent depended entirely on their own resources for starting the unit, while the remaining did approach banks, relatives, government etc. High demand, skill, ready market, better prospects were identified factors responsible for selecting a particular product. 81 percent of the respondents indicated general operational problems like managing workers, marketing and recovery of dues. 48 percent did not feel that they face problems by the virtue of being a woman.

Patel (1986) points out that initial lack of confidence in their own abilities due to a particular social backdrop further deteriorates with society's lack of confidence in women as business persons, the multiple legal, procedural and statutory formalities to be completed before and after setting up a business become a handicap for many women because of their limited mobility, also in terms of infrastructural setting. The findings show that majority have experienced discrimination when starting the business. **Young and unmarried women have been less suitable for entrepreneurial activities as the course of their own future is not in their hands.**

Rani (1986) in her study **'Potential Women Entrepreneurs'** found that the factors such as educational and income background did not influence much in women entrepreneurs' decisions to jump on the bandwagon.

Panandikar (1986) studied problems and prospects of self-employment in her paper presented at the National Seminar on Human Resources in Delhi. The paper suggested that women entrepreneurs could contribute substantially to economic development. Entrepreneurship helps women adjust her work schedule and house better.

Vinze (1987) in her research 'Women Entrepreneur in India – a socioeconomic study in Delhi' found out that women entrepreneurs opined that financial assistance from banks has been significant but procedures and formalities need to be more flexible. Management knowledge was also important and the arrangement of raw material was to be made easy, as per recommendations.

Azad (1989) in his work **'Development of Entrepreneurship among Indian Women'** states that there is a need to understand the psychological characteristics. He lists certain psychosocial factors that impede growth of women entrepreneurship: discriminating treatment from society, lack of self-confidence, poor self-image, faulty socialization, lack of encouragement, role conflict, confining to home, traditional ideology and cultural values.

Pillai (1989) states that the desire to work at the place of residence, difficulty in getting job and the desire for social recognition were the main motivating factors for self-employment. No case was found of women moving from traditional areas involving the technology and the use of science.

Anna (1990) has found that the occupational background of the father or husband provides an environment favourable to the growth of entrepreneurship among women. Author finds that highly qualified women were motivated to enter the industry in the prime of their age. However, low education did not inhibit a few entrepreneurs. The special governmental aid instilled entrepreneurial spirit.

Shah (1990) in her study **'Fostering Women Entrepreneurship – A study of distinctive features'** conducted on three income groups, it was found that entrepreneurs who had Entrepreneurship Development Program Training were sufficiently aware of the resources like knowledge of banks, state financial corporations, governmental approvals etc. It shows that from these institutions, the higher-level staff was courteous and the lower level had indifferent attitude and behaviour to accept women in the role of entrepreneurs.

Rao (1991) in his study on **'Promotion of Women Entrepreneurship'** lists economic backwardness, lack of familial and community support, ignorance of opportunities, lack of motivation, shyness and inhibition, preference for traditional occupation and liking for secure jobs as the aspects that inhibit promotion of grass root entrepreneurship among rural women.

Srivastava and Chaudhary (1991) in their work on **'Women Entrepreneurs'** found out that women faced problems mostly in marketing of their products and approaching the banks for loans. Personal problems like time constraint and family stress is also cited. Joint family is a facilitating factor and not an obstacle for women entrepreneurs; this role enhances familial bonds and increases role satisfaction of women.

Rathore and Chhabra (1991) in their work of **'Promotion of Women Entrepreneurs – Training strategies'** state that Indian women find it increasingly difficult to adjust themselves to the dual role, being tossed between home and work.

Varshneya (1991) in his study has mentioned that over past few decades there has been a substantial increase in women's participation in the economic activities all over the world. But still they receive only about 10% of world's assets, even though they represent 50% of the world's population and do $2/3^{rd}$ of the total work in the world.

Jain A. (1992) in her study **'A socioeconomic study of the artisans in Jari industry'** focused on women's role and participation in the year 1992, the result of the study indicated that all women in Jari production have taken this activity as heredity occupation.

Jyothi and Prasad (1993) in their study **'A profile of potential rural woman entrepreneur'** opine that the policy makers should also take a favourable attitude towards women entrepreneurs in granting licences, permissions, quota of scarce raw material, fair price fixation, order booking, quality testing and in the supply of technical knowledge.

Kirve and Kanitkar (1993) reveal that the awareness building, which was incorporated as a part of training input, proved extremely useful in building confidence among the trainees.

Singh (1993) in her work concluded that the factors impinging on the entrepreneurial manifestation of women are no different from those affecting men. Family and community setup did help and facilitate their entry into business. Self-fulfilment, autonomy and independence were the main reasons of their entering into profession and as such, money does not seem to be the solo motto. The researcher opines that there is no need for formal training, only proper education and adequate exposure.

Venkapathy (1993) in his work related to **'Entrepreneurial Development among Women'** finds that women with metropolitan urban background are more likely to participate in business and industry as against women with semi-urban background who are less professional, have less exposure, awareness and interest in business-related activities and generally lack management education background.

Sethi (1994) in her study **'Women Entrepreneurship in India'** lists economic compulsion, education, work experience and urbanization as the collective forces which has given and impetus to women entrepreneurship. The author says that availability of credit is the most binding of all regulatory, technical, marketing and other inputs. Although obtaining finance for the start-up is a common problem for men and women, yet these problems can be exacerbated for women because of their gender as well as prevalent socio-cultural variables. The author further finds that by and large, women are bound by their domestic and child care responsibilities and thus the choice of industry, the location and their areas of operation cannot be an individual choice. She suggests that in order to break away from the traditional women's activities like garments and food into more lucrative non-traditional fields, needs a much higher investment, the help of academic institutions, GOs and NGOs.

Tulsi et al. (1995) in their study on **'Enterprising Women'** stated that women, so far, chooses traditional fields such as dress making, embroidery, knitting, foods etc but now with the change in trend, women are venturing in computers, industrial security, stock exchange etc. Factors such as business family background, family support and encouragement, independent decision making, skills in trade, credibility, interpersonal skills, disciplined approach and self-confidence helped those women in successfully launching and managing their businesses. Non-acceptance of a woman-run business was cited as a major problem in the start-up period which was overcome.

Parimalam (1996) A large number of Indian women are slowly emerging out of a system that had oppressed and exploited them for centuries. The economic needs of the family are increasing day by day, demanding the women to take up gainful employment outside the homes. Self-employment is a challenge to women as they have to play dual roles, that of housewife, mother and business executive. Developing industrial employment opportunities and entrepreneurship among women are the need of the hour.

Rani D. L. (1996) in her studies found out that more service sector entrepreneurs were employed prior to setting up their own enterprises than women entrepreneurs in the manufacturing or trading sector. Moreover, women entrepreneurs in the service sector are more educated than in those in the trading sector. Service sectors are generally not inherited and are entirely started by the women. The ease in setting up enterprise and availability of technical know-how are considered priorities by the vocation. Manufacturing and service sector women entrepreneurs prefer expanding their business, while trading sector women entrepreneurs aspire more to setup new enterprises as they are easier to set up.

Khan (1997) in his work **'Women Entrepreneurship'** advocates the self-exploration needs to be introduced in the first phase of training program to help potential women imbibe the identity. Social responsibility should form an integral part of the module.

Singla and Syal (1997) have classified the problems faced by the women entrepreneurs into three categories i.e. Problems related to project formulation, project implementation and project operation. Group Women Entrepreneurs is suggested as a good approach as it can strengthen the individual by reinvigorating activities or skills being in danger of being exploited by vested interests. The supporting agencies should also undertake group orientation training for the members.

Verma (1998) on **'Women Entrepreneurship: A Case study of Manipur Handloom'** and found that women organised more than half of the industrial units but the major problems were lack of education and proper transport facilities.

B. Shyamala (1999) in her study "Entrepreneurship Development for Women" tries to highlight the need for women entrepreneurship, problems faced by them, existing programmes that help women to take up entrepreneurship, the institutions rendering financial support to women and suggestions regarding identified areas in which women can establish themselves as entrepreneurs. The author suggests a strategy for entrepreneurial development among women by means of providing management education as well as industrial training, enhancing indigenous techniques.

Senthil (1999) in the study "Ills plaguing Women Entrepreneurs", discusses management and managerial skill. He discusses the barriers in the way of women managers to have a successful career. These barriers are cultural, social, organizational practices, styles and legal. It is believed that they act as a major hurdle in effective development of women managers. Mostly the organizational matters decided by men are not in favour of women. Besides this, it has been noticed that there persists a wide difference in opportunities between men and women.

Rugmani Bai (1999) in her study "Entrepreneurship Development among Women -Recent Trends" discusses the entrepreneurs in the current Indian context. According to her, development of women entrepreneurship is a recent trend, it should be considered as an integral part of all development effort. The study concentrates on the functions that a woman entrepreneur needs to perform with regard to the size of business unit, exploring new prospects for venturing new business, undertaking and handling the risk and uncertainty, innovating new products, good administration and coordination skill and supervision. The author also discusses the pre-requisites for the development of women's entrepreneurial capacity.

Sundari and Geetha (2000) examined the gender disparity in access to institutional credit. In their opinion, the disparity was gradually narrowing down over the time. The empowerment of poor rural women would be possible only if they were trained and skilled for a certain employment. According to them, skill training included enterprise development, increased access to credit, new approach to markets and social, economic and political strategies.

Raghaventra (2000) has made an article, "Enterprise Development Employment Avenues for Women". He has opined in his study that enterprise development has the most existing opportunities for women who have not been able to emerge from the purdah of families, chores and responsibilities. In his study, he highlighted the factors that are responsible for a woman to be successful in any business venture.

Walokar (2001) showed in his study that most women-owned SMEs are 'single product/service' ones and no diversification have taken place. In occasional cases of diversification, the reason has been more a strategy of survival. They have only one mainline activity with low ambition for growth.

Savadi (2001) in her study, "Women entrepreneurs in India: Scope and opportunities" stated that the growth of women entrepreneurship has socio-economic significance in a country like India where economic participation of women entrepreneurs is low. They concluded that by opening a large number of Small Industrial Ventures, Women Entrepreneurs could strengthen the Industrial base, provide employment opportunities and achieve balanced regional growth. Thus, this emerging economic face of women entrepreneurs can contribute a lot to the Industrial Development of the country.

Radhakrishnan and Sellimmale (2001) made a study on, "Micro-Women entrepreneurs and socio-economic empowerment-A study". The study made an attempt to analyse the role of women in generating income through micro-entrepreneurial activities in the fishing community. It is aimed at to build relationship between women entrepreneurs and socio-economic empowerment of the community. They concluded that the micro women entrepreneurs generate more income than their made counterparts owing to the dynamism of the women in selling the captured fish in the open market. They suggested that identifying, the export market network would further boost the fisher women's economic status in the society.

Rao and Hari Narayana (2001) in his study on, "Promotion of Women Entrepreneurship" lists economic backwardness, lack of familial and community support, ignorance of opportunities, lack of motivation shyness and inhibition, preference for trade occupation and preference for secure jobs as the factors that inhibit promotion of grass root entrepreneurship among rural women.

Rani (2002) in her study, "potential women entrepreneurs" found that the desire to do something independently were the prime motivating factor to start business activity amongst the sample respondents. The women in her sample reiterated that capacity to take independent decisions of the thought of entering into entrepreneurial areas was their own and not influenced by others. The author further found that factors such as educational and income background did not influence much in their decision to jump on to entrepreneurial bandwagon.

Suguna (2002) mentioned three dimensions of women empowerment, namely economic, social and political. The economic empowerment includes skill development, income generation and credit availability, whereas the social empowerment consists of equality of treatment, equality of respect, equality of opportunity, equality of recognition and equality of status. The political empowerment represents the process of controlling power and strengthening of their vitality.

Sharma and Dhameja (2002) in their edited book **'A Majority of Women Entrepreneurs in India'** are concentrated in the light manufacturing sector (leather, garments, engineering goods, beauty products etc). The second most common category is that of services (interior design, management and placement, consultancy, nursery schools etc). This is followed by the retail trade sector with boutiques, home furnishing, automobile dealing etc. In the 1970s and 1980s women entrepreneurs were limited to 'kitchen-enterprises' – the three Ps: pickle, powder (spices) and papad – or soft, traditionally feminine enterprises, such as garments, beauty care etc. However, from the 1990s and onwards, with better levels of education, more women have chosen for entrepreneurial careers in plastics, electronics and leather related industries. However, total women entrepreneurs incline towards ventures with low investment and lesser technological barriers.

Sinha (2003) in her article, "Women entrepreneurship in the Norht-East India motivation, social support and constraints" examines the impact of motivation and social support of women entrepreneurs. According to her study, the main motivating factor to start an enterprise was to earn money by both men and women entrepreneurs.

Goyal (2004) in his study, "Women entrepreneurship and empowerment" emphasized that to ensure women entrepreneurship development, integrated efforts by various institutions and corporations are needed. So that required assistance may be provided to potential women entrepreneurs. Entrepreneurship lies at the centre of economic development. Women entrepreneurship is the process where women take lead, organize a business or Industry, and provide employment opportunities to others, though earlier women concentrated more on traditional activities due to the speed of education. Women have changed their attitudes and diverted towards non-traditional activities also.

Inbalakshmi (2004) in her article on "Development of Women Entrepreneurs" says that the present social structure has not fully accepted the idea of women as an equal partner of man. According to her the government and other voluntary organization must drawn up comprehensive plans for the development of women entrepreneurs that would ensure a bright future for the real development of women entrepreneurs.

Khan (2005) in his work on "Women Entrepreneurship Training and Small business" advocates that self-employment needs to be introduced in the first phase of training course to help potential women entrepreneurs imbibe entrepreneurial identify. Social responsibility is another important area which should form an integral part of any training module of an entrepreneurship. Capacity building, access to saving credit, organizational backing, culturally and socially acceptable men to be taken along and not to over burden men were identified as the key areas of intervention by the author for sustaining women entrepreneurs.

Pandian and Mariappan (2005) in their article on "Information technology and women entrepreneurs" highlighted the status of women entrepreneurs in developing countries. They argued that though women have little access to new technologies, there is scope to improve their lives and their families by adopting the recent advancements in science and technology. They also emphasized that information technology offered many new possibilities for women's economic, social and political empowerment.

Ghosh and Cheruvalath (2006) in their study **'Progress of Female Entrepreneurs in Low-income Countries; A Theoretical Enquiry in India'**, discussed several challenges to women entrepreneurs, including the level of education, inter-role conflicts emanating out of greater parenting responsibilities, dearth of financial assistance and socio-cultural constraints. Despite many shortcomings and constraints, they had been able to establish their own enterprises, even in a developing country like India, where the male head of the households would normally take economic decisions. The Government had emerged as a major catalyst in this area, by way of providing training and other facilities to succeed, particularly in the rural areas. With concerted efforts from all dimensions, many more women are poised to start their own business; they would leave companies to 'go alone' and would fulfil their personal dream of entrepreneurship, being their own boss and achieving success in their own business venture.

Mukherjee (2006) study **'What Motivates Women Entrepreneurs? Factors Influencing Their Motivation'**, analysed the impact of entrepreneurs characterized by competitions, financial gain and independence and the socioeconomic setting provided a base for individuals to venture into entrepreneurship. Individuals perform the entrepreneurial role because of a want to achieve, and individuals with a high requirement for achievement venture into enterprise building, whether small or large. Finally the study find on the reasons why women from the low-income group venture into business and what motivated them to do so and the analysis was conducted the two districts of Maharashtra. This study concluded that the basic factors that had a potential to influence the decisions taken by women engaged in entrepreneurial activities.

Gomathi (2006) mentioned about the problems of women entrepreneurs which are financial constraints, over dependence on intermediaries, scarcity of raw-materials, intense competition, high cost of production, low mobility, family ties, social attitudes, lack of education, lack of training, family support, practical knowledge, and self-confidence, inadequate marketing facilities and inadequate managing skills.

Antony (2007) in her study on "Women entrepreneurs on the upbeat" has taken successful women entrepreneurs in Tiruchirappalli urban limits. The study has examined the nature of business and socio-economic conditions of women entrepreneurs.

Basargekar (2007) studied **'Women Entrepreneurs: Challenges Faced'** the evolution of concepts and definitions of entrepreneurship with special reference to women. The paper provided the present scenario of women entrepreneurs in India and the various challenges, which they are facing in the changing global scenario. The study focused on marketing challenges faced by women entrepreneurs and the various strategies used by them and related organizations to overcome the challenges.

Singh R. (2008) conducted a study aimed at finding out the critical entrepreneurial competencies established women entrepreneur possess. Entrepreneurial traits like initiative, opportunity seeking, self-confidence, persuasion ability, assertiveness, influence, persistence, information, concern for high quality, commitment, sense of efficacy, systematic planning and problem solving contribute to the competency of an entrepreneur and thus affect her entrepreneurial performance

Kumari S. et. al (2010) undertook a study with an objective to study the problems of rural women entrepreneurs. The results of the study indicate lack of supportive network, financial and marketing were the major problem areas for some rural women entrepreneurs and major de - motivator for other women to initiate entrepreneurial activity.

Jubi and Chandrasekar (2010) in their study entitled **'Enterprises Development and Sustainability – Need for Entrepreneurship Education for Women'**. This paper was throwing light towards the need for women entrepreneurship as a curriculum in schools and colleges with the main objective of enterprise development and sustainability in Kerala. The study concluded that enterprise activity could be sustained if and only if the entrepreneurs were having a sound background in the different areas of entrepreneurship development. Especially this article stress on the fact that the concept of entrepreneurship must reach and every corner of the country and it must enlighten the confidence of the entrepreneurs. Once again, it must be provided with different aspects of knowledge in fulfilling the thinking capability and other relevant areas so that the organizational development can be achieved which was in turn relevant to the sustainable growth of the enterprise.

In her book **'Follow Every Rainbow', Rashmi Bansal (2013)** elaborates that the Entrepreneur maybe either an individual or a group; both are equally significant in less developed countries. Though choosing an enterprise is an individual's free choice activity, in most cases, it is the decision of the family, kin group or the community. There is need to stimulate and influence the group so that the potential entrepreneur gets the kind of encouragement and support to start and run the chosen enterprise. Women constitute a significant lot in the group.

Some studies also show that at the start-off stage, most are clustered in the service/trade sector, and only after gaining experience for 8-10 years, they exhibit entrepreneurial mobility to move to the manufacturing line.

Articles

Lalitha N. talks about women and self-employment. Majority of third world women are involved in the formal sector and economic activities in which they are disproportionately represented among the poorest of the poor. With limited education, skills and few formal employment opportunities, poor rural women in developing countries often turn to self-employment as the means of supporting themselves and informal sector is based on various types of activities like farm-based and allied activities, home-based production using non-traditional and acquired skills, retail trading etc.

Dhamija in "Women and Handicrafts: Myth and Reality" reviews about handicrafts as a means of providing income to women. In some instances, crafts are a solid source of income and can provide women with a link to their own cultural heritage. Generally the types of crafts which are introduced among women are euphemized as 'feminine crafts' because they are associated with homes.

Talmani KK in an article titled "The handicrafts industry in India" says that underlying rationale in developing village crafts in India is that they help provide employment, earn valuable foreign exchange through exports as well as preserving the cultural heritage of the nation.

A paper by **Patel, Ila** on "Women's work, Organisation and State" talks about significance of organizing women workers in the unorganized sector is acknowledged in the extensive literature on women and work in India. The paper presents conceptual perspectives of women's organisations are key to the strategy of empowering women. The strategic role of women's organisations can be seen from two perspectives – women's organisations through collective action enables women to work for change within the existing systems or transforming those structures that subordinate them and women's organisations with feminist orientation are in a unique position to give centrality to gender issues and sterile for gender equality. Through organization, poor women can utilize resources for economic activities. Support from and involvement in organisations provides women access to information and contacts.

Murthy S. in 'Women and Employment' says that the employment market has recognized its identity as well as integrity and dignity. It has certainly raised the family income and enabled them to maintain higher standard of living and create savings that could be channelized for investing into production purposes. The contribution of women in raising the level of income, output and employment and initiating a process of economic development plays a vital role in the economy.

Nagesh and Murthy study their paper **'The Effectiveness of Women Entrepreneurship Training Program: A Case Study'**. The results of the study indicate that the state and central government training centres and nodal centres, which impart training to the entrepreneurs, are successful in meeting the requirements of only two-thirds of the respondents. The reaming one-third of the respondents felt that there was no effective delivery of contents in the training programs and the same practical exposure to the contemporary business, meeting the successful and unsuccessful entrepreneurs to understand the causes of success and failures in business. The study concluded that in addition to skill development, the entrepreneurship and management development institutes should necessarily initiate steps to motive women entrepreneurs and to provide them with practical management inputs.

Rosa (2005) in an article entitled **'Women Empowerment and Family Setup'** analysed the extent of exercising the decision making power by employed women in their families and observed that the employment contributes a great degree to the empowerment of women in the family. He recommended that more opportunities should be provided to women on employment or self-employment for women empowerment in the society.

Parihar and Singh (2006) in their article explains the study on "Constraints faced by women entrepreneurs in Jammu" analysed the constraints faced by the women entrepreneurs in the state of Jammu. The study proposed that the attitude of women entrepreneurs need to be reoriented to the promotion of leadership qualities.

Rajani.N (2008) in her article, "Management training needs of women entrepreneurs" examines the quality of micro enterprise management by women in sociocultural milieu and to project the management training needs of women entrepreneurs. The study concluded that training needs are identified in the area of confidence building, competence connections and capital.

Sharma P. in her study 'Women and Economic Empowerment' suggests that the progress of women is barometers by which we can measure the common women of village in India have an uncommonly large measure of sense. Without having any degrees in economics, they can quickly sum up the feasibility and viability of a rural project. We need to involve them in our planning process in order to make programs more need based and situation specific.

Shunmugam (2011) in his article entitled **'Women's Empowerment–a Powerful Strategy for Women's Liberty and Emancipation'**, pointed out that, empowerment is control over material assets, intellectual resources and ideology. Empowerment is redistribution of power. Empowerment has become the key result to many social problems. Power may be personal and social. It is a sense of gaining control, sense of participating and a sense of decision making. The components of women empowerment are awareness building, capacity building, decision making, administering equality between men and women and so on.

Increased number of women are taking up entrepreneurship in recent times. Many significant studies have been undertaken for women entrepreneurs from economic, social and psychological perspectives. However, research on women entrepreneurs and their role as owner/ managers and employers has been largely neglected as an area of study (Patel J. R., 2005). Majority of the studies mentioned in the review of related literature have been carried out with limited purpose of covering some aspects of women in. There is a wide scope for research on women entrepreneurship in developing countries like India covering various social, psychological and economic aspects. Most of these studies do not reveal the following:

1. Outlook of husband and other family members towards and aspiring woman in business before starting her own enterprise in order to provide support, encouragement and motivation.
2. Identifying leadership qualities of women in business and how they are similar or different to men in business.
3. Impact of successful women in business on the achievement motivation of budding women entrepreneurs.
4. Understanding the use of available time at hand to balance the dual role of women at home and as entrepreneurs.

The present study, therefore, is an endeavour to identify the Achievement Motivation, Leadership and Work life Balance of Women Entrepreneurs with reference to the state of Gujarat.

CHAPTER III

RESEARCH METHODOLOGY

RESEARCH METHODOLOGY

In spite of women taking to entrepreneurship in many challenging fields, the present women entrepreneurial activity in India is not very high. There exists a definite gap between the present status of women entrepreneurial activity and their potential. This gap can be bridged by planned and all round efforts of the governmental and non-governmental agencies promoting and propagating women entrepreneurship for the all round socio-economic development.

Objectives

1. To study the socio-economic background of women entrepreneurs in Gujarat
2. To analyse motivational factors behind women entrepreneurship
3. To identify the leadership quality among women entrepreneurs in Gujarat.
4. To find out the degree of work-life balance that women entrepreneur's face and its effect on their entrepreneurial performance.
5. To suggest a framework to promote women entrepreneurship in Gujarat.

Operational Definition:

Study title - Women Entrepreneurs - A Study of Achievement Motivation, Leadership and Work-life Balance of Women Entrepreneurs of Gujarat.

Women Entrepreneurs - Women entrepreneurs represent a group of women who have broken away from the beaten track and are exploring newer avenues of economic participation. (Rani, 1996)

Achievement Motivation - The concept of achievement motivation has been used to describe individual efforts towards personal goals in social surroundings (Cassidy & Lynn, 1989). It is also found to be a typical characteristic among individuals who are successful in their work (McClelland, The Achieving Society, 1961) (McClelland, Money as motivator: some research insights., 1967).

A strong achievement motivation is probably the one characteristic most clearly describing the attitudes and the behaviour of an entrepreneur (Virtanen, 1997).

Leadership - Leadership is a procedure whereby an individual influences a group of individuals to achieve a common goal. Leadership style is characterized by the consistent patterns of exhibited behaviours, attitudes, and values. Leadership has been defined as traits, behaviours, influence, persuasion, interaction patterns, role relationships, and as administrative position; and most definitions assume that leadership involves a process whereby intentional influence is exerted by one to others in an attempt to guide relationships (Moerer-Urdahl, 2005).

Gender is essential to contemporary notions of effective styles that have transformed from a traditional masculine, autocratic style to a more feminine and androgynous style of democratic and transformational leadership (Knopik & Moerer, 2008).

Work-life Balance - Work-Life Balance does not mean an equal balance. It means the capability to schedule the hours of professional and personal life so as to lead a healthy and peaceful life. It is not a new concept. It stresses on the values, attitudes and beliefs of women regarding their age to work in organizing and balancing their work and personal life. (Deshmukh & Deshmukh, 2016)

Women turn to entrepreneurship for various reasons especially for satisfying their personal needs of self-expression an autonomy. This also makes their lives increasingly complex with pressure, intense working hours and lesser involvements in the family; hence, they juggle to satisfy both the roles. (Rani, 1996) They may receive additional help at home in the form of servants, parents, in-laws and husband but still the entire responsibility of the home rests with the woman. However, the problems remain, as the entrepreneur is not a super woman and there follows a conflict.

Relevance of this study to Social Work Profession

Social Work is an inclusive profession. Its core values hold inclusive development and acceptance of people. Women have always withstood the worst of social changes and therefore labelled as vulnerable. However, in this fast-paced world, the challenges are of a different kind. Women may have reached to the pinnacle of success but not without sacrificing their family life or other core values. It is with this intention that this study needs to take place so that the expectations, ideas and aspirations of the enterprising women of the millennium can be understood and appropriate assistance can be provided to foster their skill, capacity and talent.

Inclusion Criteria

Who is a woman entrepreneur? This question has been defined in different ways by different people at different points of time. The Government of India considered the majority participation of women in enterprises as a necessary qualification for calling such enterprises as women-enterprises. Recently, the term majority participation has been modified to majority equity participation of women in organizations. This again, could be debated as many times, women may invest but not as actively participate in the running of the enterprise. It has been thus thought that both equity participation as well as participation in the running of enterprises to a greater extent by women entrepreneurs should be a necessary criteria for calling enterprises as women-run enterprises.

Therefore, for this study, the women entrepreneurs considered were based on the following **inclusion criteria**:

1. The woman entrepreneur should have contributed to the capital of the concerned enterprises.

2. The woman entrepreneur should be actively participating in the running of the organization and in all functional areas of management.

3. The establishment of the enterprise be minimum three years.

4. The establishment must help generate income for all those employees associated with it.

Discussions with appropriate authorities at the government departments like District Industries Centres (DICs) and Centres for Entrepreneurship Development (CEDs) as well as the Chambers of Commerce and Industries and Entrepreneurship Development Institute (EDI) were done. With the preliminary information provided, an attempt to formulate a list of women-run enterprises was created. It was felt that as no census data is available on the number of women entrepreneurs in the state, **a snowball sampling method** be followed.

This study follows **an exploratory cum descriptive research design** as it will help the researcher to understand the issue more thoroughly, before attempting to quantify mass responses into statistically inferable data. Used properly, exploratory research will provide rich quality information that will help identify the main issues that should be addressed in our survey and significantly reduce the level of prejudice. Within the exploratory design, the researcher will incorporate the use of Case studies. Two case studies from each zone will be presented. Reliability on Secondary Sources of Data will also be adopted.

Tools of Data Collection:

A properly structured tool for collecting demographic data will be constructed using more of close-ended questions as the study aims at obtaining the opinions of its respondents. This tool will be developed by referring the various standardised tools available on the relevant subject. A pilot sample will be studied and the sample data will be treated on Cronbach's Alpha Test to assure of the reliability and validity of the tool.

Universe:

For the purpose of this study, the researcher will study all the women-owned, women-run enterprises within the State of Gujarat. However, four major zones i.e. Ahmedabad, Rajkot, Surat and Vadodara will be covered.

Sampling:

Snowball sampling method will be used. In this method of sampling, the existing subjects under study refer the future subjects from among their acquaintances. The research scholar will also try to obtain the names of Women Entrepreneurs from the local Chamber of Commerce and Industries from each of the zones under study and proper representation from each zone will be selected for the study. All women-owned, women-run enterprises established since the last three years in the cities of Ahmedabad, Rajkot, Surat and Vadodara will be studied.

Inclusion Criteria

1. The woman entrepreneur should have contributed to the equity capital of her enterprise.
2. The woman entrepreneur should be actively participating in the running of the organization and in all functional areas of management.
3. The establishment of the enterprise be minimum three years.
4. The establishment must help generate income for all those employees associated with it.

Based on the inclusion criteria set for the study, the researcher has been able to collect data from 158 respondents from Ahmedabad, Rajkot, Surat and Vadodara cities of the state of Gujarat, India. The bifurcation of the respondents, city-wise is as under:

S. No.	City	No. of Respondents
1.	Ahmedabad	50
2.	Rajkot	19
3.	Surat	50
4.	Vadodara	39
TOTAL		**158**

Variables:

Dependent Variables: Work-Life Balance, Motivation and Leadership

Independent Variables: Age, Education, Experience and Income

Hypotheses:

1. Higher the income, higher is the motivation to work. (H_1)

2. With increased experience in the field, entrepreneurs develop their own leadership. (H_2)

3. Work-life balance is absent in women entrepreneurs. (H_3)

Statistical Treatment:

The filled questionnaires will be numbered serially and checked for consistency of the data, and any in-consistent data. The questions and responses will be codified to represent the data in tabular / graphical form wherever appropriate. The collected data will be correlated with the various dependent and independent variables. Hypotheses will be verified using appropriate statistical tools and with the help of Excel / SPSS (Statistical Package for Social Science for Windows).

Limitations:

Every research is but mere manipulation of facts and hence may carry certain prejudices and limitations. In this research, the researcher feels that there may be certain limitations which might affect the study. These limitations are,

- Lack of willingness to participate in the research.
- Non-availability of Respondents.
- Lack of understanding about the topic in concern.
- Authenticity of the secondary sources of data.

Conclusion

In spite of women taking to entrepreneurship in many challenging fields, the present women entrepreneurial activity in India is not very high. There exists a definite gap between the present status of women entrepreneurial activity and their potential. This gap can be bridged by planned and all round efforts of the governmental and non-governmental agencies promoting and propagating women entrepreneurship for the all round socio-economic development.

High need for achievement, high need for autonomy, high need for creativity, and prevalence of initiative, industriousness, and foresight through self-reliance and possession of managerial skills are all effective pointers to an entrepreneur. (Sarada, 2004). In this globalised and digitalised era, many women have found their place in the market as entrepreneurs. With the changes in the way a business runs, it becomes important to study how motivated women are to pursue their dream of enterprising. There is a great scope to study the work-life balance and leadership skills of the women of the 21st century as this era is an era of continual transformation and change and entrepreneurs have to work up to their mark and even beyond to survive in an otherwise competitive market. The Government also has rolled out the "Start-up India" initiative to encourage more and more entrepreneurs in the various business sectors.

Social Work is an inclusive profession. Its core values speak of inclusive development and acceptance of people. Women have always withstood the worst of social changes and therefore labelled as vulnerable. However, in this fast-paced world, the challenges are of a different kind. It has been a long pending economic independence that women have achieved on their own merit. Former leaders of the country like Mahatma Gandhi and Dr. B. R. Ambedkar had always encouraged participation of women in the freedom struggle and economic development of the country. Despite this ideology, it took years together for women to achieve this freedom. In recent years, women have started being more and more assertive about their economic independence and many such women are turning to entrepreneurship. They are now their own bosses. Women may have reached to the pinnacle of success but not all success comes easy. Due to the dual social roles that women play, success would not have been possible without sacrificing their family life or other social expectations. It is with this intention that this study needs to take place so that the expectations, ideas and aspirations of the enterprising women of the millennium can be understood and appropriate assistance can be provided to foster their skill, capacity and talent.

CHAPTER IV

DATA ANALYSIS AND INTERPRETATION

DATA ANALYSIS AND INTERPRETATION

This chapter attempts to analyze the data collected by applying appropriate statistical tests and interpret the same keeping in mind broad and specific objectives of the current study of 'Women Entrepreneurship: A Study of Achievement Motivation, Leadership and Work life Balance of Women Entrepreneurs of Gujarat. Statistics like simple frequency distribution tables, Cross tables, Mean, Standard Deviation, Chi-square and Analysis of Variance (AnoVa) was applied to understand the associations between the variables.

For the multiple responses, observations and interaction during data collection the separate categories were formed from the opinions and responses received. This chapter is divided in to seven sections:

SECTION I - Primary Information/ Demographic Details of Respondents

SECTION II - Achievement Motivation of The Respondents

SECTION III - Leadership of The Respondents

SECTION IV - Work Life Balance of Respondents

SECTION V - Details of Association And Correlation Between Variables

SECTION VI – Case Studies of Selected Respondents

Before moving on further with the chapter, the researcher would like to express the details of the challenges faced during the data collection. Despite the limitations being mentioned earlier in the Research Methodology part of the Introduction Chapter, the researcher here would like to elaborate on the challenges with reference to identifying the respondents and fitting them in the study.

Challenges Faced during Data Collection

Data collection is an important stage in the entire research process. Data collection, usually occurring simultaneously with data analysis, is defined as the systematic gathering of data for a particular purpose from various sources, including, interviews, focus groups, observation, existing records, and electronic devices. During the data collection period for this study, the researcher faced many unanticipated challenges for numerous reasons.

The initial challenge posed in front of the researcher was the sampling. Once the researcher had set the inclusion criteria, it became a challenging task to identify the prospective respondents across the four cities chose for the study. While the researcher assumed the data might be easily available at the various chambers of commerce and industries of each of the cities, it turned out that except for two of the four cities, such systematic data was not available. Hence, the researcher had to adopt to a snowball sampling method to arrive at a reasonable number to study.

The next challenge was to convince the researchers to participate in the study. Since many respondents were busy and also traveling for their work purposes, it came across as a bigger challenge to reach up to their pace especially in the three cities of Ahmedabad, Rajkot and Surat. Many times, it so happened that despite seeking prior appointment, the researcher was unable to meet the respondents due to their busy schedules.

Another challenge the researcher encountered was establishing rapport and getting the respondents to participate in the study. Since the questions included mentioning their annual income and family income, many respondents showed unwillingness to participate. Moreover, the researcher had to assure them that the study, in no way, would reveal their identities publicly and the researcher will use the data for research purpose only. On the other hand, the researcher also met few respondents who were willing to know the results of the study, whenever published.

Since the researcher had not travelled extensively to the three cities, other than Vadodara, for data collection, it became a challenge to reach from one place to the other, while adhering to the appointments. Moreover, if any of the appointments was cancelled at the last minute, it became a challenge to fix an appointment with another respondent immediately. Thus, time management became challenging.

SECTION I

PRIMARY INFORMATION/ DEMOGRAPHIC DETAILS OF RESPONDENTS

Primary Information/ Demographic Details of women entrepreneurs include Personal characteristics, social characteristics, general characteristics and Economic characteristics. Personal characteristics include Age, Education, Experience and Training. Social characteristics include Marital Status, Type of Family, Nature of family etc.

The researcher has been able to collect data from 158 respondents from Ahmedabad, Rajkot, Surat and Vadodara cities of the state of Gujarat, India. The bifurcation of the respondents, city-wise is as under:

S. No.	City	No. of Respondents
1.	Ahmedabad	50
2.	Rajkot	19
3.	Surat	50
4.	Vadodara	39
TOTAL		**158**

Based on this data, further analysis was done using various statistical tools.

In this section, the demographic details of the respondents have been analyzed and graphically represented.

Table 1.1 Showing Age of the Respondents

N=158

Categories	Frequency	Percent
21-30 Yrs.	43	27.2
31-40 Yrs.	59	37.3
41-50 Yrs.	34	21.5
51 and Above	20	12.7

NR	02	1.3
Total	**158**	**100**

From the above table, it can be observed that 37% i.e., majority of the respondents fall between the age groups of 31-40 years. Almost 27% respondents fall between the age group of 21-30 years whereas 22% respondents fall within the age group of 41 - 50 years. There is a significant population of respondents, 13% within the age ranges of 51 years and above whereas 2 respondents i.e., 1% have chosen to not respond to this question.

Graph 1.1 Showing Age of Respondents

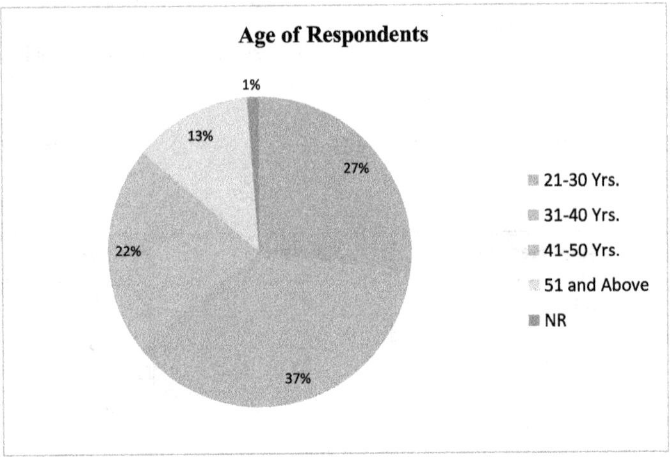

Table 1.2. Showing Educational Qualifications of the Respondents

N=158

Categories	Frequency	Percent
Secondary	07	4.4
Higher Secondary	19	12.0
Graduation	71	44.9
Post-Graduation	52	32.9
Other	07	4.4
NR	02	1.3
Total	**158**	**100**

From the above table, it can be observed that 44.9% i.e., majority of the respondents have pursued education up to Graduation level. A sizeable number of respondents, 32.9% have pursued education up to Post Graduate level whereas 12% respondents have attained education up to Higher Secondary. 4.4% respondents have pursued education until secondary level and 4.4% respondents have responded to other education received. However, 2 respondents i.e., 1.3% have chosen not to respond to this question.

Graph 1.2 Showing Educational Qualification of respondents

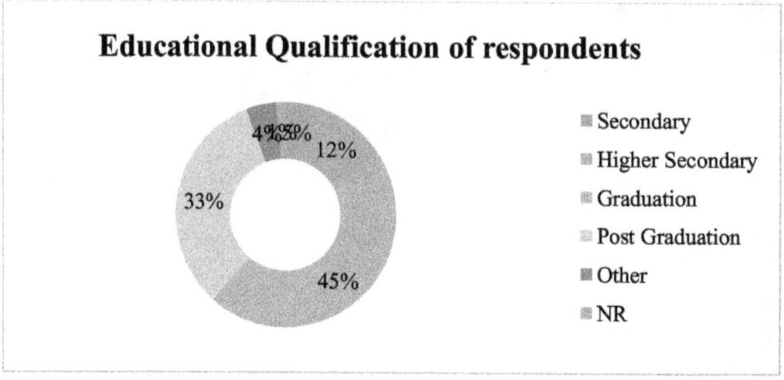

Table 1.3. Showing the Marital Status of Respondents

N=158

Categories	Frequency	Percent
Unmarried	28	17.7
Married	122	77.2
Separated/ Widow	08	5.1
Total	**158**	**100**

From the above table, it can be observed that 77.2% i.e., majority of the respondents are married. 17.7% are unmarried whereas 5.1% respondents are either separated or widowed at the time of data collection.

Graph 1.3 Showing marital status of respondents

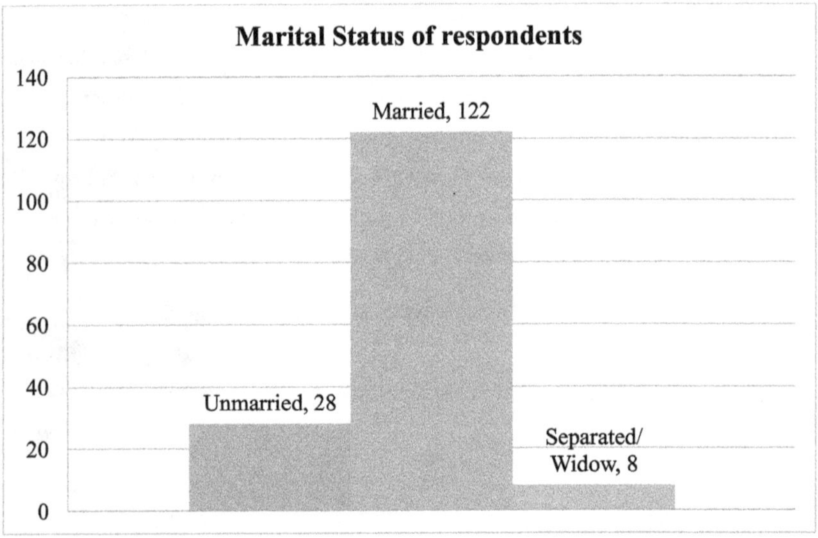

Table 1.4. Showing the experience of respondents (in years)

N=158

Categories	Frequency	Percent
<=1 Yrs	07	4.4
2-5 Yrs.	46	29.1
6-10 Yrs.	38	24.1
>10 Yrs.	54	34.2
NR	13	8.2
Total	**158**	**100**

From the above table, it can be seen that the majority of the respondents, i.e., 34.2% have an experience of more than 10 years in running an enterprise, 29.1% respondents have an experience of 2-5 years, 24.1% respondents have an experience of 6-10 years and 4.4% respondents have recently come up with their venture. In all 13 respondents chose to not respond to this question.

Graph 1.4 Showing total years of experience of respondents

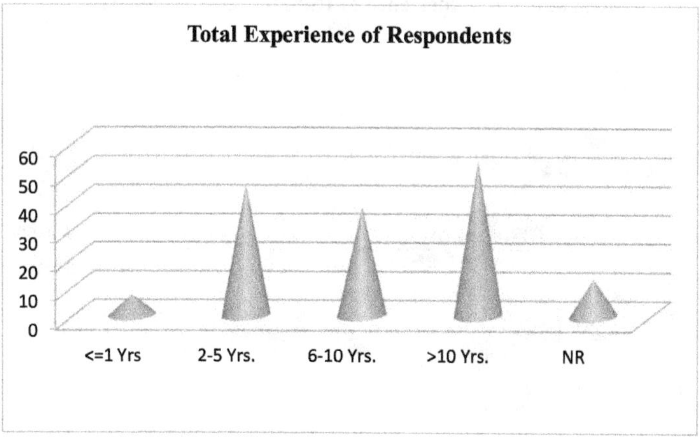

Table 1.5. Showing Experience of running current enterprise

N=158

Categories	Frequency	Percent
<=1 Yrs	20	12.7
2-5 Yrs.	60	38.0
6-10 Yrs.	38	24.1
>10 Yrs.	31	19.6
NR	9	5.7
Total	**158**	**100**

From the above table, it can be seen that the majority of the respondents, i.e., 38% have an experience of 2-5 years in running their current enterprise, 24.1% respondents have an experience of 6-10 years of running their current enterprise, 19.6% respondents have an experience of more than 10 years and 12.7% respondents have recently come up with their own venture. In all 09 respondents chose to not respond to this question.

Graph 1.5 Showing respondents' experience of running own enterprise

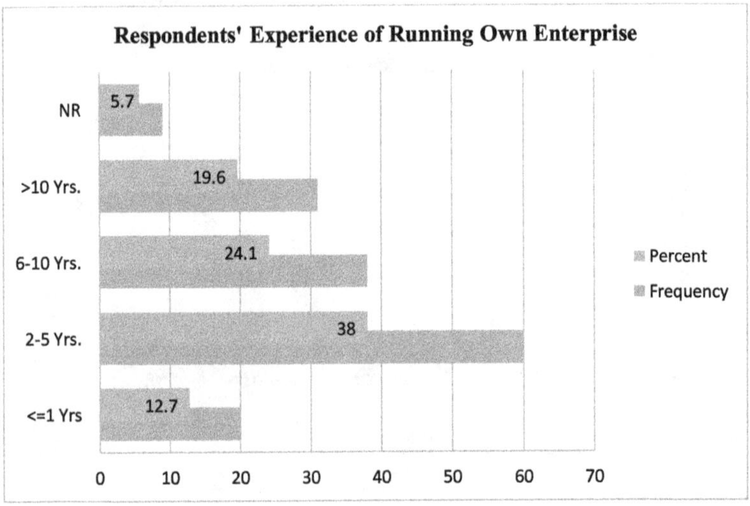

Table 1.6. Showing average monthly income of the respondents

N=158

Categories	Frequency	Percent
<= Rs.10,000 Rs.	25	15.8
Rs. 10,001-20,000	26	16.5
Rs. 21001-50,000	41	25.9
> Rs.50,000	34	21.5
Not Regular	32	20.3
Total	**158**	**100.0**

From the above table, it can be seen that the majority of the respondents, i.e., 25.9% have an average monthly income of Rs. 21,001 - 50,000, 21.5% respondents have an average income of more than Rs. 50,000, 16.5% respondents have an average income of Rs. 10,001 - 20,000 while 15.8% respondents have an average income of less than Rs. 10,000. In all 32 respondents i.e. 20.3% chose 'Not Regular" as their response.

Graph 1.6 Showing Average Monthly Income of Respondents

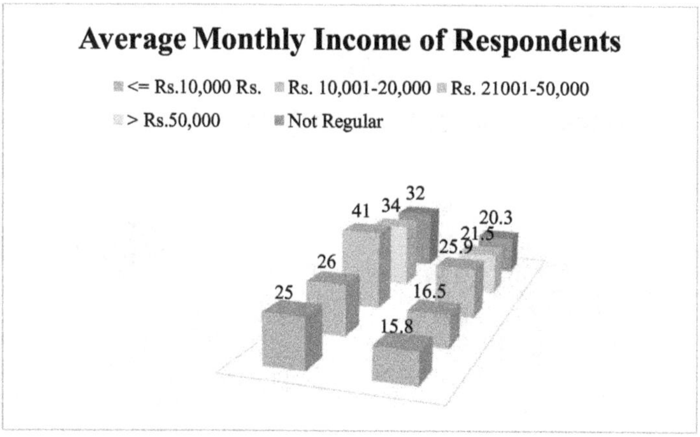

Table 1.7. Showing husband's occupation

N=158

Category	Frequency	Percent
Job	28	17.7
Business	43	27.2
Professional	43	27.2
NR	08	5.1
NA	36	22.8
Total	**158**	**100**

From the above table, it can be seen that the occupation of majority of the respondents' husbands is business i.e. 27.2% and an equal percent belong to professional category. 17.7% of the respondents answered that their husbands were doing a job while 5.1% respondents refused to respond to this question. Since there were 36 respondents who were not married, this question is not applicable to them.

Graph 1.7 Showing Occupation of Respondents' Husbands

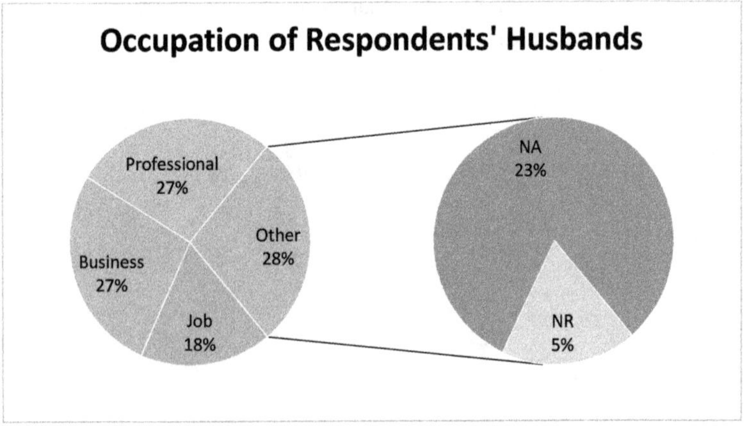

able 1.8. Showing Husband's Monthly Income

N=158

Category	Frequency	Percent
<= Rs. 10,000	04	2.5
Rs.10,001-20,000	08	5.1
Rs.21001-50,000	31	19.6
> Rs. 50,000	39	24.7
NR	40	25.3
NA	36	22.8
Total	**158**	**100**

From the above table, it can be seen that the majority of the respondents, i.e., 25.3% have chosen not to respond to this question of their husband's income, while 24.7% respondents have claimed that the average monthly income of their husbands is more than Rs. 50,000/- while 19.6% women have responded their husbands' incomes to be between Rs. 21,001 to Rs. 50,000/-; for 5.1% females, the average monthly income of their husbands is between Rs. 10,001 - Rs. 20,000/- while for just 2.5% respondents, the average monthly income of their husbands is less than Rs. 10,000/-. It can be seen that 22.8% women were not married at the time of data collection and hence, this question is not applicable to them.

Graph 1.8 Showing Average Monthly Income of Respondents' Husbands

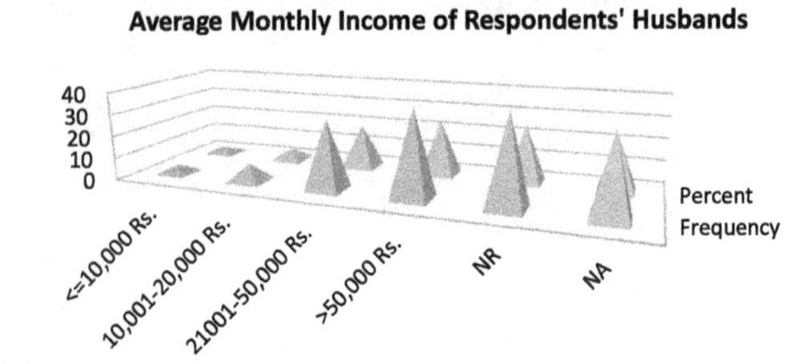

	<=10,000 Rs.	10,001-20,000 Rs.	21001-50,000 Rs.	>50,000 Rs.	NR	NA
Frequency	4	8	31	39	40	36
Percent	2.5	5.1	19.6	24.7	25.3	22.8

Table1.9. Showing Husband's Educational Qualification

N=158

Categories	Frequency	Percent
<=12th Std.	15	9.5
Graduation	57	36.1
Post-Graduation	24	15.2
Professional	19	12.0
NR	7	4.4
NA	36	22.8
Total	**158**	**100.0**

From the above table, it can be seen that the majority of the respondents, i.e., 36% have husbands whose minimum qualification is graduation, while 15% respondents have claimed that th0 minimum educational qualification of their husbands is post graduation. Husbands of 12% respondents have professional degrees like CA, MBA, Medical Professionals etc. and 10% women's' husbands are only 12th pass. 4% respondents have chosen not to answer this question while for 23% respondents, this question is not applicable.

Graph 1.9 Showing Respondent's Husband's Educational Qualification

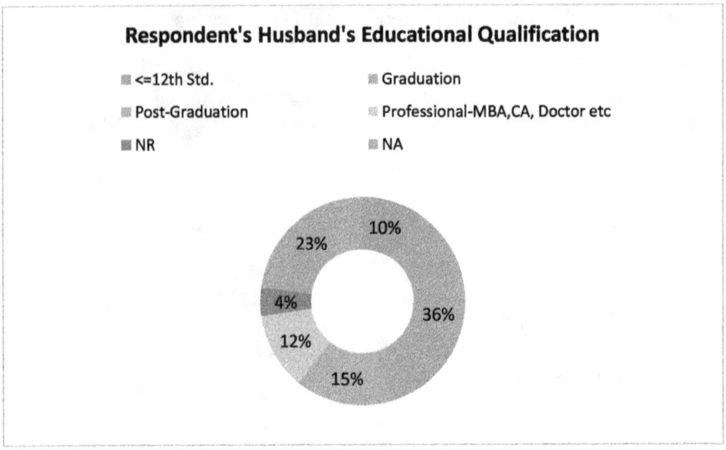
Respondent's Husband's Educational Qualification
<=12th Std.
Graduation
Post-Graduation
Professional-MBA,CA, Doctor etc
NR
NA
10%
23%
4%
36%
12%
15%

Table 1.10 Showing number of Family Members

N=158

Categories	Frequency	Percent
1-2 members	10	6.3
3-5 members	129	81.6
>=6 members	17	10.8
NR	02	1.3
Total	**158**	**100**

From the above table, it can be seen that the majority of the respondents, i.e., 81.6% have 3-5 members in their families, while 10.8% respondents have more than 6 members in their families. 6.3% respondents have 1-2 members in their families whereas 1.3% respondents have not responded to this question.

Graph 1.10 Showing Number of Family Members of Respondents

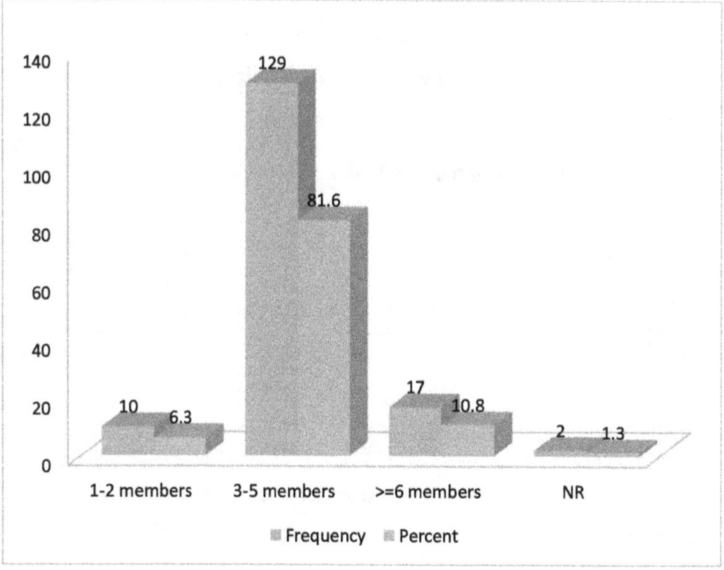

Table 1.11 Showing number of children of the respondents

N=158

Category	Frequency	Percent
0	14	8.9
1	51	32.3
2	50	31.6
3	07	4.4
4	03	1.9
NR	05	3.2
NA	28	17.7
Total	**158**	**100**

From the above table, it can be seen that the majority of the respondents, i.e., 32.3% have only one child while 31.6% respondents have 2 children in their families. Almost 4.4% respondents have 3 children while 1.9% respondents have 4 children. Almost 8.9% respondents have no children while 3.2 respondents have not responded to this question. This question is not applicable to 17.7% respondents.

Graph 1.11 Showing Number of Children of Respondents

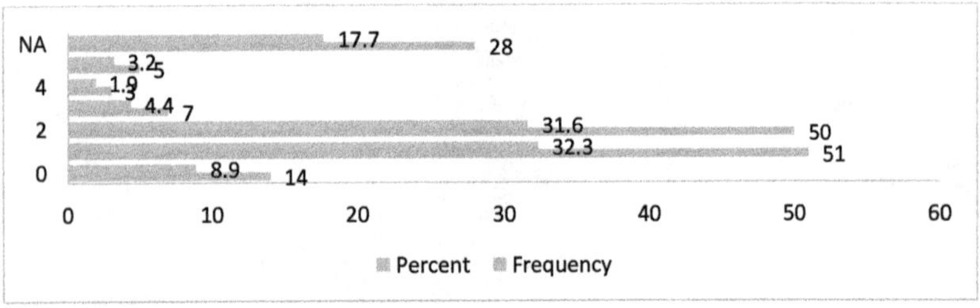

The following chapter goes on to discuss the associations between the independent and dependent variables of the current study.

SECTION II

RESULTS ON ACHIEVEMENT MOTIVATION OF RESPONDENTS

Achievement Motivation is defined as behaviors dedicated to developing and demonstrating higher abilities. It is one of the three components that make up McClelland's Human Motivation Theory. Those who are motivated by achievement tend to have a particular set of features. The first is that they choose their battles wisely. They are not gamblers when it comes to setting goals and will not set one that is impossible. That being said, they will not be as motivated by a task that can be easily accomplished either.

The respondents were asked to fill in their level of achievement motivation and their responses were further analyzed for correlation.

Table 2.1 Showing Achievement Motivation of Respondents

N=158

Particulars	Completely Disagree (CD)		Mostly Disagree (MD)		Agree to some extent (ASE)		Mostly Agree (MA)		Completely Agree (CA)		Total	
	F	%	F	%	F	%	F	%	F	%	F	%
I like reading the biography of great people in order to learn how they overcame hurdles and achieved great things in life	11	7.00	11	07	49	31	45	28.5	42	26.6	**158**	**100**

I plan ahead what to do during my free time	08	5.10	14	8.9	30	19	50	31.6	56	35.4	**158**	**100**
When I come to know that somebody like me, have achieved something great, I am motivated to do the something in a better way.	06	3.80	08	5.1	19	12	51	32.3	74	46.8	**158**	**100**
Most people who know me say that I am hard working and ambitious.	07	4.40	09	5.7	35	22.2	49	31	58	36.7	**158**	**100**
I never leave a task/assignment, I start unfinished	04	2.50	12	7.6	29	18.4	57	36.1	56	35.4	**158**	**100**
I always work very hard to be among the best.	11	7.00	06	3.8	17	10.8	50	31.6	74	46.8	**158**	**100**
I aim at reaching the highest level at my work.	11	7.00	08	5.1	28	17.7	38	24.1	73	46.2	**158**	**100**
I want to do something	08	5.10	11	07	34	21.5	35	22.2	70	44.3	**158**	**100**

which others have not done.												
I am basically a competitive person and I compete just for the sake of competing.	30	19.00	24	15.2	46	29.1	35	22.2	23	14.6	**158**	**100**
I believe that success in life has less to do with hard work and more to do with luck and being in the right place at the right time.	35	22.20	26	16.5	39	24.7	39	24.7	19	12	**158**	**100**
I will go ahead with my plans only if I am sure that other people will approve of it.	39	24.70	31	19.6	36	22.8	30	19	22	13.9	**158**	**100**
It is not a good idea to be always above others in achievement, because that may make them feel bad about themselves.	31	19.60	34	21.5	41	25.9	37	23.4	15	9.5	**158**	**100**
I like to be the best.	10	6.30	9	5.7	30	19	38	24.1	71	44.9	**158**	**100**

I enjoy finishing my assignments even when they are difficult and time consuming.	11	7.00	08	5.1	31	19.6	41	25.9	67	42.4	**158**	**100**
I enjoy making friends with the most intelligent workers so as to keep up my standards of performances.	21	13.30	08	5.1	34	21.5	46	29.1	49	31	**158**	**100**
I like when people say in front of others that I am doing well.	08	5.10	15	9.5	32	20.3	44	27.8	59	37.3	**158**	**100**
I would like to deal with difficult situations, so that the blame or praise for its results come to me alone.	17	10.80	25	15.8	35	22.2	36	22.8	45	28.5	**158**	**100**

The above table shows the respondents' level of achievement motivation. From the above table, a cumulative Achievement Motivation score is obtained. Since there was no greater variance between the values of Completely Disagree (CD) and Mostly Disagree (MD) as well as Mostly Agree (MA) and Completely Agree (CA), the researcher has clubbed the items in the cumulative table that can be represented as below:

Table 2.1.1 Cumulative response on Achievement Motivation

N=158

Achievement Motivation		
	Frequency	**Percent**
Mostly disagree	15	9.5%
Agree to some extent	90	57%
Mostly agree	53	33.5%
Total	**158**	**100%**

The above graph shows the cumulative responses of the respondents on Achievement Motivation. It shows that out if the 158 respondents, majority of the respondents, i.e., 57% agree to some extent on having Achievement Motivation, 33.5% mostly agree that they have high Achievement Motivation and 9.5% of respondents disagree on having Achievement Motivation.

Graph 2.1.1 Showing Achievement Motivation of Respondents

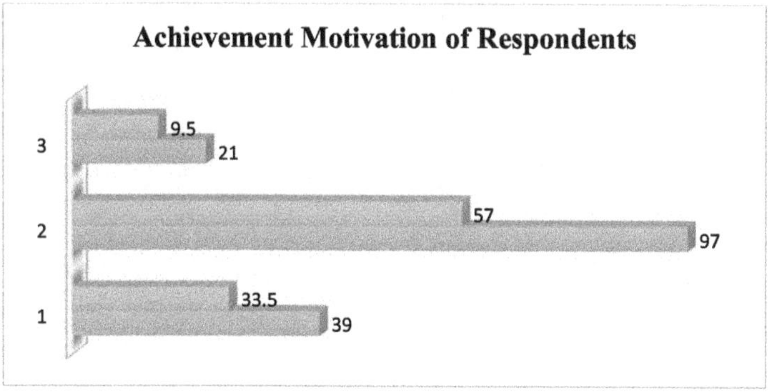

Thus, it can be concluded that there is visible level of achievement motivation present among the respondents. It is this motivation that aspires women to work harder and achieve greater heights in their career. There is a great sense of achievement in women, and on knowing that people like them have achieved much in life, they are also motivated to work harder.

This also hints at the aspiring and hardworking nature of the women respondents.

Respondents own their actions and believe that they themselves like to deal with difficult situations so that the blame or praise comes to them alone.

SECTION III

RESULTS ON LEADERSHIP OF RESPONDENTS

Leadership is an important virtue for entrepreneurs, as it will help the entrepreneurs to hold their workforce together during every stage of the development of business. A good leader will be able to bring along all the people along with him/ her whereas a leader who cannot lead his/ her people well will gradually end up losing resources. For an entrepreneur, resources are of prime concern. It becomes the duty of an entrepreneur to also be a good leader who believes in the potential of the people who work with them.

The respondents were asked to fill in their level of leadership and their responses were further analyzed for correlation.

Table 3.1 Showing Responses of Respondents on Leadership

N=158

Items	**Never**		**Sometimes**		**Always**		**Total**	
	F	**%**	**F**	**%**	**F**	**%**	**F**	**%**
I encourage my team to participate when it comes decision-making time and I try to implement their ideas and suggestions.	11	7%	55	34.80 %	92	58.20 %	**158**	**100 %**
Nothing is more important than accomplishing a goal or task.	13	8.20 %	44	27.80 %	101	63.90 %	**158**	**100 %**

I closely monitor the schedule to ensure a task or project will be completed in time.	05	3.20 %	39	24.70 %	114	72.20 %	**158**	**100 %**
I enjoy coaching people on new tasks and procedures.	07	4.40 %	60	38%	91	57.60 %	**158**	**100 %**
The more challenging a task is, the more I enjoy it.	15	9.50 %	52	32.90 %	91	57.60 %	**158**	**100 %**
I encourage my employees to be creative about their job.	10	6.30 %	48	30.40 %	100	63.30 %	**158**	**100 %**
When seeing a complex task through to completion, I ensure that every detail is accounted for.	10	6.30 %	54	34.20 %	94	59.50 %	**158**	**100 %**
I find it easy to carry out several complicated tasks at the same time.	27	17.10 %	73	46.20 %	58	36.70 %	**158**	**100 %**

I enjoy reading articles, books, and journals about training, leadership, and psychology; and then putting what I have read into action.	31	19.60 %	70	44.30 %	57	36.10 %	**158**	**100 %**
When correcting mistakes, I do not worry about jeopardizing relationships.	26	16.50 %	83	52.50 %	49	31%	**158**	**100 %**
I manage my time very efficiently.	10	6.30 %	53	33.50 %	95	60.10 %	**158**	**100 %**
I enjoy explaining the intricacies and details of a complex task or project to my employees.	19	12%	61	38.60 %	78	49.40 %	**158**	**100 %**
Breaking large projects into small manageable tasks is second nature to me.	17	10.80 %	65	41.1%^	76	48.10 %	**158**	**100 %**
Nothing is more important than	09	5.70 %	59	37.30 %	90	57%	**158**	**100 %**

building a great team.								
I enjoy analyzing problems.	13	8.20 %	55	34.80 %	90	57%	**158**	**100 %**
I honour other people's boundaries.	10	6.30 %	48	30.40 %	100	63.30 %	**158**	**100 %**
Counselling my employees to improve their performance or behaviour is second nature to me.	15	9.50 %	59	37.30 %	84	53.20 %	**158**	**100 %**
I enjoy reading articles, books, and trade journals about my profession; and then implementing the new procedures I have learned.	21	13.30 %	60	38%	77	48.70 %	**158**	**100 %**

The above table shows the respondents' level of Leadership. From the above table, a cumulative Leadership score is obtained. The researcher has clubbed the items in the cumulative table that can be represented as below:

Table 3.1.1 Showing Cumulative responses on Leadership

N=158

	Frequency	Percent
Sometimes	39	24.7
Always	119	75.3
Total	**158**	**100.0**

Since none of the responses hint towards negation in Leadership, the 'Never' option has been deleted for easy understanding, The above table can be graphically represented as:

Graph 3.1.1 Showing Cumulative responses on Leadership.

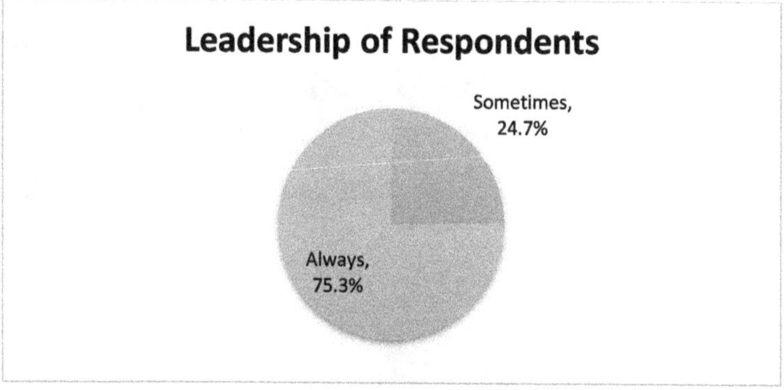

From the data, it can be said that most of the women accept that they closely monitor the tasks at hand.

Many respondents believe in understanding and respecting other peoples' boundaries and limitations thereby knowing how to delegate tasks to them. This also makes them efficient leaders as they understand and accept their team members as they are. A more androgynous cum transformational leadership style is visible among women entrepreneurs.

Respondents feel they need to be more connected with their staff and hence may overlook mistakes however, if the mistakes can lead to serious consequences, they do not hesitate in putting their personal relations apart from their professional ones.

SECTION IV

RESULTS ON WORK LIFE BALANCE OF RESPONDENTS

Work-Life Balance does not mean an equal balance. It means the capacity to schedule the hours of professional and personal life so as to lead a healthy and peaceful life. It is not a new concept. It emphasizes the values, attitudes and beliefs of women regarding their age to work in organizing and balancing their work and personal life. (Deshmukh & Deshmukh, 2016)

The respondents were asked to fill in their level of Work Ife Balance and their responses were further analyzed for correlation.

Table 4.1 Showing Work life Balance of Respondents

N=158

Items	Not True		Little True		Somewhat True		Fairly True		Definitely True		NA		Total	
	F	%	F	%	F	%	F	%	F	%	F	%	F	%
I do not find it difficult to take leave at the time of social emergencies.	20	12.7%	10	6.30%	38	24.10%	56	35.40%	34	21.50%	00	0.00%	158	100%
I am able to participate in community	09	5.70%	14	8.90%	31	19.60%	62	39.20%	42	26.60%	00	0.00%	158	100%

activities and attend to religious commitments.														
I do not find enough time to spend with my family and friends.	50	31.60%	30	19%	35	22.20%	28	17.70%	15	9.50%	00	0.00%	158	100%
I comfortably fulfil the basic requirement of my family.	13	8.20%	07	4.40%	34	21.50%	32	20.30%	72	45.60%	00	0.00%	158	100%
I help my children in preparing for	09	5.70%	13	8.20%	14	8.90%	37	23.40%	47	29.70%	38	24.10%	158	100%

their exams.														
I do not get time for my sick partner / child/parents.	69	43.70%	23	14.60%	33	20.90%	19	12%	14	8.90%	00	0.00%	158	100%
I get the opportunity to enjoy holidays with my family.	06	3.80%	14	8.90%	38	24.10%	43	27.20%	57	36.10%	00	0.00%	158	100%
I get time to invite my friends for a party at home.	18	11.40%	21	13.30%	34	21.50%	34	21.50%	51	32.30%	00	0.00%	158	100%
I comfortably fulfil the basic require	12	7.60%	07	4.40%	27	17.10%	42	26.60%	70	44.30%	00	0.00%	158	100%

ment of my family.														
I meet prescribed deadlines and schedules, without affecting my home life.	08	5.10%	12	7.6^	32	20.30%	49	31.10%	57	36.10%	00	0.00%	158	100%
I experience work pressure while doing a group task.	25	15.80%	25	15.80%	36	22.80%	43	27.20%	29	18.40%	00	0.00%	158	100%
I work for extra hours to get my work done.	12	7.60%	10	6.30%	38	24.10%	44	27.80%	54	34.20%	00	0.00%	158	100%
I often take additional	32	20.30%	20	12.70%	33	20.90%	31	19.60%	42	26.60%	00	0.00%	158	100%

work home.													
I feel pressure while working when given a deadline.	22	13.90%	21	13.30%	39	24.70%	43	27.20%	33	20.90%	00	0.00%	158	100%
I share the work with my colleagues whenever needed.	17	10.80%	21	13.30%	30	19%	46	29.10%	43	27.20%	01	6%	158	100%
I cannot manage more than one project at a time.	51	32.30%	27	17.10%	27	17.10%	27	17.10%	26	16.50%	00	0.00%	158	100%
I do not do overtime to comple	48	30.40%	22	13.90%	30	19%	31	19.60%	27	17.10%	00	0.00%	158	100%

te my work.														
I can adjust my working schedule to attend to my life priorities.	04	2.50%	19	12%	33	20.90%	47	29.70%	55	34.80%	00	0.00%	158	100%
I do exercises and take care of my health.	24	15.20%	21	13.30%	28	17.70%	41	25.90%	44	27.80%	00	0.00%	158	100%
I enjoy doing my job.	05	3.20%	08	5.10%	12	7.60%	30	19%	103	65.20%	00	0.00%	158	100%

The above table shows the respondents' level of Work life Balance. From the above table, a cumulative Work life Balance score is obtained. The researcher has clubbed the items in the cumulative table that can be represented as below:

Table 4.1.1 Showing cumulative responses of Work life Balance of Respondents

N=158

	Frequency	Percent
If it little true	01	0.6
If it is somewhat true	39	24.7
If it is fairly true	97	61.4
If it is definitely true	21	13.3
Total	**158**	**100.0**

The above table shows that out of the total respondents, i.e. 158, for a majority of respondents, 61.4% achieving work life balance is fairly true, for 24.7% respondents, achieving work life balance is somewhat true, for 13.3% respondents, achieving work life balance is definitely true whereas for only a small population of 0.6%, it is difficult to achieve work life balance.

This can be more descriptively represented through a graph, as below:

Graph 4.1.1 Showing cumulative responses of Work life Balance of Respondents

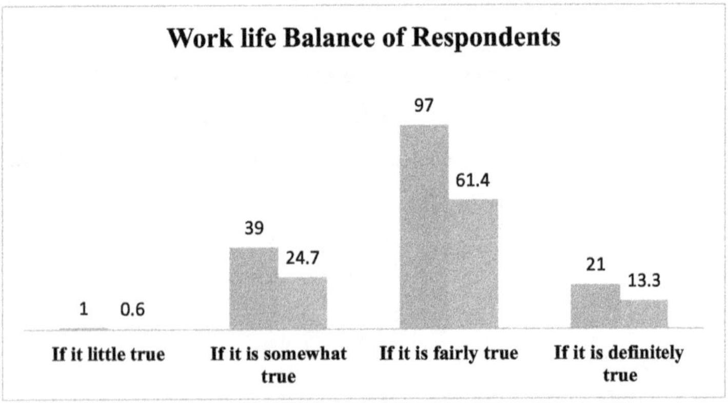

As far as Work-life Balance is concerned, the respondents felt that their personal lives were not affected by their work and they could easily balance it out.

Respondents expressed that they felt a deep sense of satisfaction by contributing to their family and the society at large by being in business and this feeling motivated them to push harder to achieve their goals.

SECTION V

DETAILS OF ASSOCIATION AND CORRELATION BETWEEN VARIABLES

Table 5.1 Showing Association between Age and Achievement Motivation

N=158

			Age of the Respondent				**Total**	**Pearson Chi-Square Value**	**df**	**Sig.**
			21-30 Yrs.	31-40 Yrs.	41-50 Yrs.	51 and Above				
Achievement Motivation	**MD**	Count	0	13	2	0	15	21.578[a]	6	**0.001**
		% within Age of the Respondent	0.00%	22.00%	5.90%	0.00%	9.50%			
	ASE	Count	29	32	19	10	90			
		% within Age of the Respondent	65.90%	54.20%	55.90%	47.60%	57.00%			
	MA	Count	15	14	13	11	53			
		% within Age of the Respondent	34.10%	23.70%	38.20%	52.40%	33.50%			
Total		**Count**	**44**	**59**	**34**	**21**	**158**			

	% within Age of the Respondent	**100.00%**	**100.00%**	**100.00%**	**100.00%**	**100.00%**			

(The mean difference is significant at the 0.05 level.)

(MD = Mostly Disagree; ASE = Agree to Some Extent; MA = Mostly Agree)

The Above table represents the association between Age, an Independent Variable and Achievement Motivation, a Dependent Variable. The Chi-Square Value is 21.578[a] with significance level 0.001. It can therefore be interpreted that Achievement Motivation is highly significant with Age. This means as respondents age increases, their motivation to work on their business and flourish it, also increases.

Table 5.2 Showing Association between Education and Achievement Motivation

N=158

			Education					**Total**	**Chi-Square Value**	**df**	**Sig.**
			SSC	**HSC**	**Graduation**	**PG**	**Other**				
Achievement Motivation	**MD**	Count	0	6	5	4	0	15	13.630[a]	8	0.092
		% within Education	0.00%	31.60%	6.80%	7.70%	0.00%	9.50%			
	ASE	Count	4	8	44	29	5	90			
		% within Education	57.10%	42.10%	60.30%	55.80%	71.40%	57.00%			
		Count	3	5	24	19	2	53			

	MA	% within Education	42.90%	26.30%	32.90%	36.50%	28.60%	33.50%			
Total		**Count**	**7**	**19**	**73**	**52**	**7**	**158**			
		% within Education	**100.00%**	**100.00%**	**100.00%**	**100.00%**	**100.00%**	**100.00%**			

(The mean difference is significant at the 0.05 level.)

(MD = Mostly Disagree; ASE = Agree to Some Extent; MA = Mostly Agree)

The above table shows the association between Education, an Independent Variable and Achievement Motivation, a Dependent Variable of the current study. The table does not show any significant association between the Independent and Dependent Variables (p-value=0.092). This clearly emphasizes that Achievement Motivation does not significantly depend on Education.

Table 5.3 Showing Association between Experience and Achievement Motivation

N=158

			Experience				**Total**	**Chi-Square Values**	**df**	**Sig.**
			<=1 Yrs	**2-5 Yrs.**	**6-10 Yrs.**	**>10 Yrs.**				
Achievement Motivation	**MD**	Count	0	4	11	0	15	20.806[a]	6	**0.002**
		% within Experience	0.00%	6.60%	24.40%	0.00%	9.50%			
	ASE	Count	15	37	23	15	90			
		% within Experience	68.20%	60.70%	51.10%	50.00%	57.00%			
	MA	Count	7	20	11	15	53			

		% within Experience	31.80%	32.80%	24.40%	50.00%	33.50%			
Total		**Count**	**22**	**61**	**45**	**30**	**158**			
		% within Experience	**100.00%**	**100.00%**	**100.00%**	**100.00%**	**100.00%**			

(The mean difference is significant at the 0.05 level.)

(MD = Mostly Disagree; ASE = Agree to Some Extent; MA = Mostly Agree)

The above table attempts to show a association between Experience, an Independent Variable and Achievement Motivation, a Dependent Variable. The Chi-Square Value for association is 20.806[a] and the Significant p-Value is 0.002.

Since the mean difference is considered significant at 0.05 level and the significance value for this correlation is much less, the above variables are found to be highly associated with each other.

Hence, Achievement Motivation is highly correlated with Experience, which signifies that women entrepreneurs tend to gain more Achievement Motivation with increased experience in the field.

Table 5.4 Showing Association between Average Income (Per Month) and Achievement Motivation

N=158

			Average Income (Per Month)					**Total**	**Chi-Square Value**	**df**	**Sig.**
			<= Rs.10,000	**Rs. 10,001-20,000**	**Rs. 21001-50,000**	**> Rs. 50,000**	**Not Regular**				
		Count	1	0	3	3	8	15		10	

	MD	% within Average Income (Per Month)	4.00%	0.00%	7.30%	8.80%	25.00%	9.50%			
Achievement Motivation	ASE	Count	19	17	20	18	16	90	20.699[a]		**0.023**
		% within Average Income (Per Month)	76.00%	65.40%	48.80%	52.90%	50.00%	57.00%			
	MA	Count	5	9	18	13	8	53			
		% within Average Income (Per Month)	20.00%	34.60%	43.90%	38.20%	25.00%	33.50%			
Total		**Count**	**25**	**26**	**41**	**34**	**32**	**158**			
		% within Average Income (Per Month)	**100.00%**	**100.00%**	**100.00%**	**100.00%**	**100.00%**	**100.00%**			

(The mean difference is significant at the 0.05 level.)

(MD = Mostly Disagree; ASE = Agree to Some Extent; MA = Mostly Agree)

The above table attempts to show an association between Average Income (Per Month), an Independent Variable and Achievement Motivation, a Dependent Variable. The Chi-Square Value for association is 20.699[a] and the significant p-Value is 0.023.

Since the mean difference is considered significant at 0.05 level and the significance value for this correlation is much less, the above variables are found to be highly correlated with each other.

It can be inferred from this data set that higher income develops greater achievement motivation among women entrepreneurs. Hence, from the available results, the hypothesis for this study - Higher the Income, Higher the Motivation to work, is accepted.

Table 5.5 Showing association between Age and Leadership

N=158

			Age of the Respondent					Chi-Square Value	df	Sig.
			21-30 Yrs.	**31-40 Yrs.**	**41-50 Yrs.**	**51 and Above**	**Total**			
Leadership	**Sometimes**	Count	8	17	9	5	39	1.609[a]	3	0.657
		% within Age	18.20%	28.80%	26.50%	23.80%	24.70%			
	Always	Count	36	42	25	16	119			
		% within Age	81.80%	71.20%	73.50%	76.20%	75.30%			
Total		**Count**	**44**	**59**	**34**	**21**	**158**			
		% within Age	**100.00%**	**100.00%**	**100.00%**	**100.00%**	**100.00%**			

(The mean difference is significant at the 0.05 level.)

The above table shows association between Age, and Independent Variable and Leadership, a Dependent Variable. The Chi-Square Value is 1.609[a] and significant p-Value is 0.657. Since the mean difference is significantly accepted at 0.05 level and the obtained score is 0.657, the variables are not significantly correlated with each other.

Hence, Age is not associated with Leadership.

Table 5.6 Showing association between Education and Achievement Motivation

N=158

			Education						Chi-Square Value	df	Sig.
			SSC	**HSC**	**Graduation**	**PG**	**Other**	**Total**			
Leadership	**Sometimes**	Count	3	4	21	8	3	39	5.695[a]	4	0.223
		% within Education	42.90%	21.10%	28.80%	15.40%	42.90%	24.70%			
	Always	Count	4	15	52	44	4	119			
		% within Education	57.10%	78.90%	71.20%	84.60%	57.10%	75.30%			
Total		**Count**	**7**	**19**	**73**	**52**	**7**	**158**			
		% within Education	**100.00%**	**100.00%**	**100.00%**	**100.00%**	**100.00%**	**100.00%**			

(The mean difference is significant at the 0.05 level.)

The above table shows the association between Education, and Independent Variable and Leadership, a Dependent Variable. The Chi-square Value is 5.695[a] and the significance is 0.223. Since the mean difference is significantly accepted at 0.05 level and the obtained score is 0.223, the variables are not significantly correlated with each other.

Hence, Education is not associated with Leadership

Table 5.7 Showing association between Experience and Leadership

N=158

			Experience					Chi-Square Value	df	Sig.
			<=1 Yrs	2-5 Yrs.	6-10 Yrs	>10 Yrs	Total			
Leadership	Sometimes	Count	5	12	18	4	39	8.627[a]	3	**0.035**
		% within Experience	22.70%	19.70%	40.00%	13.30%	24.70%			
	Always	Count	17	49	27	26	119			
		% within Experience	77.30%	80.30%	60.00%	86.70%	75.30%			
Total		**Count**	**22**	**61**	**45**	**30**	**158**			
		% within Experience	**100.00%**	**100.00%**	**100.00%**	**100.00%**	**100.00%**			

(The mean difference is significant at the 0.05 level.)

The above table shows association between Experience, an Independent Variable and Leadership, a Dependent Variable. The Chi-Square Value is 8.627[a] and Significant p-Value is 0.035. Since the mean difference is significantly accepted at 0.05 level and the obtained score is 0.035, the variables are highly significant and correlated with each other.

From the obtained data for the current study, The hypothesis (H_2) that - With increased Experience, Women Entrepreneurs develop their own Leadership is accepted within the purview of this study.

Table 5.8 Showing association between Income and Leadership

N=158

			Average Income (Per Month)					**Total**	**Chi-Square Value**	**df**	**Sig.**
			<= Rs. 10,000	**Rs. 10,001- 20,000**	**Rs. 21001- 50,000**	**> Rs. 50,000**	**Not Regular**				
Leadership	**Sometimes**	Count	8	12	7	5	7	39	10.950[a]	5	0.052
		% within Average Income (Per Month)	32.00%	46.20%	17.10%	14.70%	21.88%	24.70%			
	Always	Count	17	14	34	29	25	119			
		% within Average Income (Per Month)	68.00%	53.80%	82.90%	85.30%	78.13%	75.30%			
Total		Count	**25**	**26**	**41**	**34**	**32**	**158**			
		% within Average Income (Per Month)	**100.00%**	**100.00%**	**100.00%**	**100.00%**	**100.00%**	**100.00%**			

(The mean difference is significant at the 0.05 level.)

The above table shows association between Average Income (Per Month), an Independent Variable and Leadership, a Dependent Variable. The Chi-Square Value is 10.950[a] and Significant p-Value is 0.052. Since the mean difference is significantly accepted at 0.05 level and the obtained score is 0.052, the variables are partially significant and correlated with each other.

It can be inferred from the obtained data that with an increase in the Income, the leadership is also enhanced. Hence, Income and Leadership are partially significant with each other.

Table 5.9 Showing association between Age and Work life Balance

N=158

			Age of the Respondent				**Total**	**Value**	**df**	**Sig.**
			21-30 Yrs.	**31-40 Yrs.**	**41-50 Yrs.**	**51 and Above**				
Work Life Balance	**if it little true**	Count	0	0	1	0	1	13.367[a]	9	0.147
		% within Age of the Respondent	0.00%	0.00%	2.90%	0.00%	0.60%			
	if it is some what true	Count	16	15	7	1	39			
		% within Age of the Respondent	36.40%	25.40%	20.60%	4.80%	24.70%			
	if it is fairly true	Count	22	38	22	15	97			
		% within Age of the Respondent	50.00%	64.40%	64.70%	71.40%	61.40%			
	if it is definitely true	Count	6	6	4	5	21			
		% within Age of the Respondent	13.60%	10.20%	11.80%	23.80%	13.30%			
Total		**Count**	**44**	**59**	**34**	**21**	**158**			
		% within Age of the Respondent	**100.00%**	**100.00%**	**100.00%**	**100.00%**	**100.00%**			

(The mean difference is significant at the 0.05 level.)

The above table shows association between Age, an Independent Variable and Work life Balance, a Dependent Variable. The Chi-Square Value is 13.367[a] and Significant p-Value is 0.147. Since the mean difference is significantly accepted at 0.05 level and the obtained score is 0.147, the variables are not significantly correlated with each other.

It can be inferred from the obtained data that Work life balance of Women Entrepreneurs has no association with their Age. Hence, Age and Work Life Balance are not significant with each other.

Table 5.10 Showing association between Education and Work life Balance

N=158

			Education					Total	Chi-Square Value	df	Sig.
			SSC	HSC	Graduation	PG	Other				
Work Life Balance	if it little true	Count	0	0	1	0	0	1	10.316[a]	12	0.588
		% within Education	0.00%	0.00%	1.40%	0.00%	0.00%	0.60%			
	if it is some what true	Count	4	4	19	9	3	39			
		% within Education	57.10%	21.10%	26.00%	17.30%	42.90%	24.70%			
	if it is fairly true	Count	3	13	41	36	4	97			
		% within Education	42.90%	68.40%	56.20%	69.20%	57.10%	61.40%			
	if it is definitely true	Count	0	2	12	7	0	21			
		% within Education	0.00%	10.50%	16.40%	13.50%	0.00%	13.30%			

Total	**Count**	**7**	**19**	**73**	**52**	**7**	**158**			
	% within Education	**100.00%**	**100.00%**	**100.00%**	**100.00%**	**100.00%**	**100.00%**			

(The mean difference is significant at the 0.05 level.)

The above table shows association between Education, an Independent Variable and Work life Balance, a Dependent Variable. The Chi-Square Value is 10.316[a] and Significant p-Value is 0.558. Since the mean difference is significantly accepted at 0.05 level and the obtained score is 0.558, the variables are not significantly associated with each other.

It can be inferred from the obtained data that Work life balance of Women Entrepreneurs has no association with their Education. Hence, Education and Work Life Balance are not significantly associated with each other.

Table 5.11 Showing association between Experience and Work life Balance

N=158

			Experience				**Total**	**Chi-Square Value**	**df**	**Sig.**
			<=1 Yrs	**2-5 Yrs.**	**6-10 Yrs.**	**>10 Yrs.**				
Work Life Balance	**if it little true**	Count	0	1	0	0	1	15.587[a]	9	0.076
		% within Experience	0.00%	1.60%	0.00%	0.00%	0.60%			
	if it is some what true	Count	9	14	15	1	39			
		% within Experience	40.90%	23.00%	33.30%	3.30%	24.70%			
		Count	10	37	27	23	97			

	if it is fairly true	% within Experience	45.50%	60.70%	60.00%	76.70%	61.40%			
	if it is definitely true	Count	3	9	3	6	21			
		% within Experience	13.60%	14.80%	6.70%	20.00%	13.30%			
Total		**Count**	**22**	**61**	**45**	**30**	**158**			
		% within Experience	**100.00%**	**100.00%**	**100.00%**	**100.00%**	**100.00%**			

(The mean difference is significant at the 0.05 level.)

The above table shows association between Experience, an Independent Variable and Work life Balance, a Dependent Variable. The Chi-Square Value is 15.587[a] and Significant p-Value is 0.076. Since the mean difference is significantly accepted at 0.05 level and the obtained score is 0.076, the variables are not significantly associated with each other.

It can be interpreted from the obtained data that Work life balance of Women Entrepreneurs has no association with their Experience in the field. Hence, Experience and Work Life Balance are not significantly associated with each other.

Table 5.12 Showing association between Income and Work Life Balance

N=158

			Average Income (Per Month)								
			<=Rs. 10,000	Rs.10,001-20,000	Rs.21001-50,000	> Rs. 50,000	Not Regular	Total	Chi-Square Value	df	Sig.
Work Life Balance	if it little true	Count	0	0	0	0	1	1	114.149[a]	15	0.00
		% within Avg. Income (Per Month)	0.00%	0.00%	0.00%	0.00%	3.13%	0.60%			
	if it is somewhat true	Count	15	8	2	4	10	39			
		% within Avg. Income (Per Month)	60.00%	30.80%	4.90%	11.80%	31.25%	24.70%			
	if it is fairly true	Count	9	16	32	21	19	97			
		% within Avg. Income (Per Month)	36.00%	61.50%	78.00%	61.80%	59.38%	61.40%			
	if it is definitely true	Count	1	2	7	9	2	21			
		% within Avg. Income (Per	4.00%	7.70%	17.10%	26.50%	6.25%	13.30%			

		Month)									
Total		Count	25	26	41	34	32	158			
		% within Avg. Income (Per Month)	100.00%	100.00%	100.00 %	100.00%	100.00%	100.00%			

(The mean difference is significant at the 0.05 level.)

The above table shows association between Average Income (Per Month), an Independent Variable and Work life Balance, a Dependent Variable. The Chi-Square Value is 114.149[a] and Significant p-Value is 0.000. Since the mean difference is significantly accepted at 0.05 level and the obtained score is 0.000, the variables are highly significant and associated with each other.

It can be inferred from the obtained data that Work life balance of Women Entrepreneurs has significant correlation with their Income.

From the obtained data for the current study, the hypothesis formulated that - Work life Balance is absent among Women Entrepreneurs (H_3), is not accepted. This concludes that despite other factors not letting respondents achieve a balance between their work life and their family life, higher income helps them to do so.

Hence, Income and Work Life Balance are significantly associated with each other.

On the basis of the data collected and statistically treated statistical, a collective result on the overall association between Achievement Motivation, Leadership and Work life Balance of Respondents was prepared.

A tabular representation of the same is as below:

Table 5.13 Showing Association between Independent and Dependent Variables.

N=158

		Achievement Motivation	**Leadership**	**Work Life Balance**	**Age of the Respondent**	**Education**	**Experience**	**Average Income (Per**
Achievement Motivation	Pearson Correlation	01	.456**	.367**	.095	.055	.001	.003
	Sig. (2-tailed)		**.000**	**.000**	.235	.493	.994	.974
	N	158	158	158	158	158	158	126
Leadership	Pearson Correlation	.456**	01	.422**	.011	.077	.062	.301**
	Sig. (2-tailed)	**.000**		**.000**	.890	.334	.441	**.001**
	N	158	158	158	158	158	158	126
Work Life Balance	Pearson Correlation	.367**	.422**	01	.197*	.028	.168*	.376**

	Sig. (2-tailed)	**.000**	**.000**		**.013**	.723	**.035**	**.000**
	N	158	158	158	158	158	158	126

(The mean difference is significant at the 0.05 level.)

The above table shows association between the independent and dependent variables. It can be seen that Achievement Motivation is highly significant and associated with Leadership and Work life Balance whereas, it is not significant with any of the Independent Variables.

Leadership is highly significant and associated with the dependent variable of Work life balance and the Independent variable of Average Income (Per Month). It can be interpreted that respondents who have higher income have better leadership skills. However, there is no significance between leadership and experience. Hence, the hypothesis that with greater experience, the respondents develop better leadership skills is rejected. This gives scope to study the leadership patterns of respondents and what factors foster leadership among women entrepreneurs.

Work life Balance is highly significant with Age of the respondents, Experience of running own enterprise and Average Income (per month). This finding emphasizes on the fact that with increased age, respondents try to find their own ways to achieve work life balance. Hence the hypothesis, Work life Balance is absent in Women Entrepreneurs (H_3) cannot be out rightly rejected as during their initial day of start-up, respondents do find it difficult to achieve work life balance, hence it is absent, however, gradually, respondents learn to strike a balance between their work life and their family life.

Work life Balance is also greatly significant with Experience of running own enterprise. With increased experience of running own enterprises, respondents find out ways how they can balance their work life and their personal life.

Work life balance is also highly significant with Average Monthly Income. This shows that as respondents start earning more, they are able to exhibit better leadership in terms of delegating and managing their work in such a manner that they can achieve better work life balance.

SECTION VI

REPRESENTATION OF *f*-TEST (ANOVA) TABLES

On the basis of the exposed data to the statistical tests, a collective result on the overall correlation between Achievement Motivation, Leadership and Work life Balance of Respondents along with each Independent variable was prepared. A tabular representation of the same is given below.

Table 6.1 Showing correlation between Age and Achievement Motivation, Leadership and Work life balance

N=158

		N	**Mean**	**Std. Deviation**	**df**	**F**	**Sig**
Achievement Motivation	21-30 Yrs.	44	63.48	7.914	3	**4.566**	**.004**
	31-40 Yrs.	59	58.08	12.381			
	41-50 Yrs.	34	63.21	9.422			
	51 and Above	21	66.05	7.221			
	Total	**158**	61.75	10.387			
Leadership	21-30 Yrs.	44	62.34	9.175	3	**0.409**	**.747**
	31-40 Yrs.	59	60.44	13.491			
	41-50 Yrs.	34	60.68	14.968			
	51 and Above	21	63.52	14.562			
	Total	**158**	61.43	12.873			
	21-30 Yrs.	44	46.75	10.486	3	**2.691**	**.048**

Work Life Balance	31-40 Yrs.	59	48.07	9.557			
	41-50 Yrs.	34	48.62	11.404			
	51 and Above	21	54.19	8.334			
	Total	**158**	48.63	10.261			

(The mean difference is significant at the 0.05 level.)

From the table, it can be seen that by comparing average scores by Age groups of the respondents Achievement Motivation and work life Balance are significant (p-values=.004 and 0.048) where as it is not significant for the Leadership.

Therefore, it can be interpreted that there is greater significance between Age and Achievement Motivation and between Age and Work life Balance. But no significance is seen between Age and Leadership. This brings us to an understanding that both Achievement Motivation and Work life Balance increase with age.

Here, the hypothesis that Work life Balance is absent among women entrepreneurs is accepted in the sense that at an earlier age women entrepreneurs find it difficult to manage their work life and family life, but with age, they are able to strike balance between their work life and family life.

Since there is an evident correlation between Age and Achievement Motivation and Age and Work life Balance, Post Hoc test was carried out to find the details of correlation. The data exposed to Post Hoc test shows the following result:

Table 6.1.1 Showing Multiple Comparisons of Age and Achievement Motivation

Scheffe				
Dependent Variable	**(I) Age of the Respondent**	**(J) Age of the Respondent**	**Std. Error**	**Sig.**
Achievement Motivation	31-40 Yrs.	21-30 Yrs.	2.002	.068
		41-50 Yrs.	2.164	.138
		51 and Above	2.554	.024

From the above table, it can be interpreted that there is greater significance of Achievement Motivation within the age group of 31-40 years (p-value=0.024) in multiple comparison with respondents of 51 years and above.

Table 6.2 Showing correlation between Educational Qualification and Achievement Motivation, Leadership and Work life Balance

N=158

		N	Mean	Std. Deviation	df	F	Sig.
Achievement Motivation	Secondary	7	66.29	6.525	4	**1.809**	**0.130**
	Higher Secondary	19	56.26	14.247			
	Graduation	73	62.19	9.683			
	Post-Graduation	52	62.56	10.065			
	Other	7	61.43	7.613			
	Total	158	61.75	10.387			
Leadership	Secondary	7	57.43	13.049	4	**1.464**	**0.216**
	Higher Secondary	19	61.79	11.048			
	Graduation	73	60.01	13.129			
	Post-Graduation	52	64.54	12.642			
	Other	7	56.14	14.588			
	Total	158	61.43	12.873			
Work Life Balance	Secondary	7	43.14	11.006	4	**0.893**	**0.469**
	Higher Secondary	19	48.95	9.210			

		N	Mean	Std. Deviation	df	F	Sig.
	Graduation	73	49.08	11.039			
	Post-Graduation	52	49.21	9.392			
	Other	7	44.29	10.226			
	Total	158	48.63	10.261			

(The mean difference is significant at the 0.05 level.)

From the table, it can be seen that by comparing average scores by Educational Qualifications of respondents Achievement Motivation, Leadership and Work life Balance, no significant difference is found (p-values=0.130, p-value=0.216 and p-value=0.469 respectively) Therefore, it can be interpreted that there is no significance between Educational Qualification and Achievement Motivation, Leadership and Work life Balance of the respondents.

Table 6.3 Showing correlation between Marital Status and Achievement Motivation, Leadership and Work life Balance

N=158

		N	Mean	Std. Deviation	df	F	Sig.
Achievement Motivation	Unmarried	28	65.43	7.686	2	**2.938**	**0.056**
	Married	122	60.67	10.716			
	Separated/Widow	8	65.25	10.899			
	Total	**158**	61.75	10.387			
Leadership	Unmarried	28	61.57	10.031	2	**0.514**	**0.599**
	Married	122	61.11	13.256			
	Separated/Widow	8	65.88	16.331			

		N	Mean	Std. Deviation	df	F	Sig.
	Total	**158**	61.43	12.873			
Work Life Balance	Unmarried	28	47.89	12.020	2	**0.306**	**0.737**
	Married	122	48.64	9.652			
	Separated/Widow	8	51.13	13.590			
	Total	**158**	48.63	10.261			

(The mean difference is significant at the 0.05 level.)

From the table, it can be seen that by comparing average scores by Marital Status of the respondents Achievement Motivation is partially significant (p-value=.056) where as it is not significant for Leadership (p-value=0.599) and Work life Balance (p-value=0.737).

Therefore, it can be interpreted that there is little significance between Marital Status and Achievement Motivation but no significant difference due to respondents' marital status in Leadership and Work life Balance.

Table 6.4 Showing correlation between Number of Years of Experience and Achievement Motivation, Leadership and Work life Balance

N=158

		N	Mean	Std. Deviation	df	F	Sig.
Achievement Motivation	<=1 Yrs.	7	64.00	5.657	3	**1.228**	**0.301**
	2-5 Yrs.	49	63.00	9.760			
	6-10 Yrs.	39	59.10	11.159			
	>10 Yrs.	63	62.16	10.662			
	Total	**158**	61.75	10.387			
Leadership	<=1 Yrs.	7	62.29	7.111	3	**0.578**	**0.630**

	2-5 Yrs.	49	60.55	10.275			
	6-10 Yrs.	39	59.87	12.620			
	>10 Yrs.	63	62.98	15.189			
	Total	**158**	61.43	12.873			
Work Life Balance	<=1 Yrs.	7	48.71	13.647	3	**0.613**	**0.607**
	2-5 Yrs.	49	47.39	10.679			
	6-10 Yrs.	39	48.08	10.484			
	>10 Yrs.	63	49.94	9.476			
	Total	**158**	48.63	10.261			

(The mean difference is significant at the 0.05 level.)

From the table, it can be seen that by comparing average scores on Number of Years of Experience and Achievement Motivation, Leadership and Work life Balance, no significant difference is observed (p-values=0.301, 0.630 and 0.607). Therefore, it can be interpreted that there is no greater significance between Number of Years of Experience and Achievement Motivation, Leadership and Work life Balance of the respondents.

Table 6.5 Showing correlation between Experience of running own enterprise and Achievement Motivation, Leadership and Work life Balance

N=158

		N	**Mean**	**Std. Deviation**	**df**	**F**	**Sig.**
Achievement Motivation	<=1 Yrs.	22	63.23	7.709	3	5.892	**.001**
	2-5 Yrs.	61	62.75	9.025			
	6-10 Yrs.	45	56.82	12.826			

	>10 Yrs.	30	66.00	7.926			
	Total	**158**	61.75	10.387			
Leadership	<=1 Yrs.	22	60.09	8.933	3	5.075	**.002**
	2-5 Yrs.	61	63.11	9.759			
	6-10 Yrs.	45	56.18	15.464			
	>10 Yrs.	30	66.87	14.098			
	Total	**158**	61.43	12.873			
Work Life Balance	<=1 Yrs.	22	46.50	10.949	3	3.657	**.014**
	2-5 Yrs.	61	48.46	10.048			
	6-10 Yrs.	45	46.49	10.494			
	>10 Yrs.	30	53.77	8.328			
	Total	**158**	48.63	10.261			

(The mean difference is significant at the 0.05 level.)

From the table, it can be seen that by comparing average scores on Number of Years of Experience of running own enterprise with Achievement Motivation, Leadership and Work life Balance, greater significant difference is observed (p-values=0.001, 0.002 and 0.014).

Therefore, it can be interpreted that there is greater significance between Number of Years of Experience of running own enterprise and Achievement Motivation, Leadership and Work life Balance. It can be interpreted that respondents have exhibited greater skills of leadership in running their own enterprise; they have also managed to balance their work life and family life and also maintained greater work life balance.

To further analyze the correlation between the significance of variables, the data was exposed to Post-Hoc Test, which used the Scheffe method to determine the comparisons and revealed the following results:

Table 6.5.1 Showing Multiple Comparisons for Experience and Achievement Motivation, Leadership and Work life Balance

Scheffe					
Dependent Variable	**(I) Experience**	**(J) Experience**	**Mean Difference (I-J)**	**Std. Error**	**Sig.**
Achievement Motivation	6-10 Yrs.	2-5 Yrs.	-5.932*	1.952	**0.029**
		>10 Yrs.	-9.178*	2.341	**0.002**
Leadership	6-10 Yrs.	2-5 Yrs.	-6.937*	2.437	**0.048**
		>10 Yrs.	-10.689*	2.923	**0.005**
Work Life Balance	6-10 Yrs.	>10 Yrs.	-7.278*	2.359	**0.026**

(The mean difference is significant at the 0.05 level.)

It can be seen from the above table that by applying Post-Hoc Studies using the Scheffe's method it is found that Achievement Motivation is highly significant with respondents having 6-10 years of experience with respondents having 2-5 years of experience (p-value=0.029) as well as those having more than 10 years of experience (p-value=0.002).

Similarly, Leadership is highly significant with respondents having 6-10 years of experience with respondents having 2-5 years of experience (p-value=0.048) as well as those having more than 10 years of experience (p-value=0.005).

And Work life Balance is highly significant with respondents having 6-10 years of experience with respondents having more than 10 years of experience (p-value=0.026). Thereby signifying that with more number of years in their own business, women adapt to achieving more Work life Balance.

Table 6.6 Showing correlation between Average Monthly Income and Achievement Motivation, Leadership and Work life Balance

N=126

		N	**Mean**	**Std. Deviation**	**df**	**F**	**Sig.**
Achievement Motivation	<= Rs. 10,000	25	61.8	7.427	3	0.413	**0.744**
	Rs.10,001-20,000	26	64.38	8.242			
	Rs. 21001-50,000	41	63.37	10.297			
	> Rs.50,000	34	62.24	10.552			
	Total	126	62.96	9.402			
Leadership	<= Rs. 10,000	25	55.48	9.426	3	4.218	**0.007**
	Rs.10,001-20,000	26	59.65	10.654			
	Rs. 21001-50,000	41	63.8	11.367			
	> Rs.50,000	34	65.59	14.496			
	Total	126	61.78	12.278			
Work Life Balance	<= Rs. 10,000	25	41.84	8.375	3	9.025	**0.000**

	Rs.10,001-20,000	26	46.35	10.457			
	Rs. 21001-50,000	41	52.9	8.111			
	> Rs.50,000	34	51.53	10.067			
	Total	126	48.98	10.08			

(The mean difference is significant at the 0.05 level.)

From the table, it can be seen that by comparing average scores of Average Monthly Income on Achievement Motivation there is no greater significant difference (p-value = 0.744). On comparing scores of Average Monthly Income and Leadership, greater significant difference is observed (p-value = 0.007) whereas on comparison of average scores of Average Monthly Income and Work life Balance, greater significant difference (p-value = 0.000). Therefore, it can be interpreted that there is greater significance between Average Monthly Income of respondents and Leadership and Work life Balance. However, no significant difference is seen between Achievement Motivation and Average Monthly Income of respondents.

To further analyze the correlation between the significance of variables, the data was exposed to Post-Hoc Test which used the Scheffe method to determine the comparisons and revealed the following results:

Table 6.6.1 Showing Multiple Comparisons for Average Income (Per Month) and Leadership and Work life Balance

Scheffe					
Dependent Variable	**(I) Average Income (Per Month)**	**(J) Average Income (Per Month)**	**Mean Difference (I-J)**	**Std. Error**	**Sig.**
Leadership	<= Rs. 10,000	Rs. 21001-50,000	-8.325	3.002	**0.058***
		> Rs.50,000	-10.108*	3.117	**0.017***
Work Life Balance	<=Rs. 10,000	Rs. 21001-50,000	-11.062*	2.342	**0.000***
		> Rs. 50,000	-9.689*	2.432	**0.002***
	Rs. 10,001-20,000	Rs. 21001-50,000	-6.556	2.314	**0.05***
**The mean difference is significant at the 0.05 level.*					

It can be seen from the above table that, by applying Post-Hoc Studies using the Scheffe's method, it is found that Leadership is highly significant with respondents having Average Income (Per Month) Rs <=10,000 with respondents having Average Income (Per Month) between Rs. 21,000 - 50,000 (p-value=0.058) as well as those having Average Income (Per Month) more than Rs. 50,000 (p-value=0.017).

Similarly, Work life Balance is highly significant with respondents having Average Income (Per Month) between Rs. 21,000 - 50,000 (p-value=0.000) and those having Average Income (Per Month) more than Rs. 50,000 (p-value=0.002). Also, significance is found between respondents having Average Income (Per Month) between Rs. 10,001 - 20,000 and Rs. 21,001 - Rs. 50,000 (p-value=0.005).

Here the N=126 as 32 respondents chose not to respond to this question.

Table 6.7 Showing correlation between Spouse's Occupation and Achievement Motivation, Leadership and Work life Balance

N=114

		N	Mean	Std. Deviation	df	F	Sig.
Achievement Motivation	Job	28	56.68	13.679	2	3.35	**0.039**
	Business	43	63.05	8.516			
	Professional	43	61.67	9.601			
	Total	114	60.96	10.598			
Leadership	Job	28	57.68	13.787	2	2.066	**0.132**
	Business	43	62.67	13.114			
	Professional	43	63.7	11.403			
	Total	114	61.83	12.788			
Work Life Balance	Job	28	48.46	10.976	2	0.087	**0.917**
	Business	43	49.05	7.961			
	Professional	43	49.4	9.137			
	Total	114	49.04	9.138			

(The mean difference is significant at the 0.05 level.)

From the table, it can be seen that by comparing average scores of spouse's occupation and Achievement Motivation there is greater significant difference (p-value = 0.039). On comparison of average scores by Spouse's Occupation and Leadership, greater significant difference is observed for leadership (p-value = 0.132). However, no significant difference is seen for work life balance (p-value = 0.917) and Spouse's Occupation. Therefore, it can be interpreted that there is greater significance between Spouse's Occupation of respondents and Achievement Motivation and Leadership. However, no significant difference is seen between Work life Balance and Spouse's Occupation of respondents.

To further analyze the correlation between the significance of variables, the data was exposed to Post-Hoc Test which used the Scheffe method to determine the comparisons and revealed the following results:

Table 6.7.1 Showing Multiple Comparisons of Husband's Occupation and Achievement Motivation

N=114

Scheffe				
Dependent Variable	**(I) Husband's Occupation**	**(J) Husband's Occupation**	**Mean Difference (I-J)**	**Sig.**
Achievement Motivation	Job	Business	-6.368*	0.045
*. The mean difference is significant at the 0.05 level.				

It can be seen from the above table that by applying Post-Hoc Studies using the Scheffe's method it is found that Achievement Motivation is highly significant with respondents' husbands Occupation as Job with those respondents whose husbands are doing business (p-value=0.045).

Here. N=114 as 37 respondents are not married while 07 respondents chose not to answer the question.

Table 6.8 Showing correlation between Husband's Average Monthly Income and Achievement Motivation, Leadership and Work life Balance

N=82

		N	Mean	Std. Deviation	df	F	Sig.
Achievement Motivation	<=Rs. 10,000	4	58.75	5.252	3	0.438	0.726
	Rs. 10,001-20,000	8	62.25	9.347			
	Rs.21001-50,000	31	63.58	8.168			
	> Rs.50,000	39	61.77	10.191			
	Total	82	62.35	9.132			
Leadership	<=Rs.10,000.	4	51.75	10.874	3	1.029	0.385
	Rs.10,001-20,000	8	59.88	8.593			
	Rs.21001-50,000	31	63.13	12.619			
	>Rs. 50,000	39	62.33	13.536			
	Total	82	61.88	12.722			
Work Life Balance	<=Rs. 10,000	4	40.5	3.416	3	4.213	0.008
	Rs.10,001-20,000	8	43.13	8.442			
	Rs.21001-50,000	31	51.45	8.278			
	>Rs. 50,000	39	51	8.288			
	Total	82	49.89	8.637			

(The mean difference is significant at the 0.05 level.)

From the table, it can be seen that by comparing average scores of spouse's average monthly income on Achievement Motivation there is no significant difference (p-value = 0.726). No significant difference is seen on comparing the average scores of Spouse's Average Monthly income and Leadership (p-value = 0.385). Greater significant difference is observed on comparing average scores for Work life Balance (p-value = 0.008) and Husband's Average Monthly Income. Therefore, it can be interpreted that there is greater significance between Husband's Average Monthly Income and Work life Balance. However, no significant difference is seen between Achievement Motivation and Leadership, and Husband's Average Monthly Income.

To further analyze the correlation between the significance of variables, the data was exposed to Post-Hoc Test which used the Scheffe method to determine the comparisons and revealed the following results:

Table 6.8.1 Showing multiple comparisons between Husband's Income (Per Month) and Work life Balance

N=82

Scheffe				
Dependent Variable	**(I) Husband's Income (Per Month)**	**(J) Husband's Income (Per Month)**	**Mean Difference (I-J)**	**Sig.**
Work Life Balance	Rs. 10,001-20,000	Rs. 21001-50,000	3.238	0.094
*The mean difference is significant at the 0.05 level.				

From the above table, it can be seen that there is there is partial significance within the variables of Husband's Income (Per Month) is Rs. 10,001 - 20,000 and between Rs. 21,001 - 50,000 (p-value=0.094) within the dependent variable of Work life Balance.

Here, N=82 as again, 37 respondents are not married while 39 other chose not to respond to this question.

Table 6.9 Showing correlation between Number of Family Members and Achievement Motivation, Leadership and Work life Balance

N=158

		N	**Mean**	**Std. Deviation**	**df**	**F**	**Sig.**
Achivement Motivation	1-2 members	10	60.1	3.665	3	0.503	**0.681**
	3-5 members	129	61.93	10.94			
	>=6 members	17	60.47	8.959			
	NR	2	69	7.071			
	Total	158	61.75	10.387			
Leadership	1-2 members	10	49.7	12.91	3	3.621	**0.015**
	3-5 members	129	61.91	12.005			
	>=6 members	17	63.35	16.336			
	NR	2	72.5	9.192			
	Total	158	61.43	12.873			
Work Life Balance	1-2 members	10	41.6	9.559	3	1.759	**0.157**
	3-5 members	129	48.98	10.288			

	>=6 members	17	50.12	9.937			
	NR	2	49	5.657			
	Total	158	48.63	10.261			

(The mean difference is significant at the 0.05 level.)

From the table, it can be seen that by comparing the average scores of Number of Family Members and Achievement Motivation, there is no significant difference (p-value = 0.681). Greater significant difference is observed when the comparison of average scores is done for leadership (p-value = 0.015) and Number of Family Members. No significant difference is also seen when average scores are compared with work life balance (p-value = 0.157) and Number of Family Members. Therefore, it can be interpreted that there is greater significance between Number of Family Members and Leadership. However, no significant difference is seen between Achievement Motivation and Work life Balance and Number of Family Members.

To further analyze the correlation between the significance of variables, the data was exposed to Post-Hoc Test which used the Scheffe method to determine the comparisons and revealed the following results:

Table 6.9.1 Showing multiple comparisons of No. of Family Members and Leadership

N=158

Scheffe				
Dependent Variable	**(I) No. of Family Members**	**(J) No. of Family Members**	**Mean Difference (I-J)**	**Sig.**
Leadership	1-2 members	3-5 members	-12.215*	0.036
**The mean difference is significant at the 0.05 level.*				

It can be seen from the above table that Leadership is highly significant between those respondents who have 1-2 family members with those who have 3-5 family members (p-value=0.036).

Table 6.10 Showing correlation between Husband's Educational Qualification and Achievement Motivation, Leadership and Work life Balance

N=115

		N	Mean	Std. Deviation	df	F	Sig.
Achievement Motivation	<=12th Std.	16	61.63	10.158	2	3.966	**0.022**
	Graduation	54	63.35	9.688			
	Post Graduate/ Post Professional-MBA, CA, Doctor etc	45	57.31	11.992			
	Total	115	60.75	10.986			
Leadership	<=12th Std.	16	63.25	13.473	2	0.191	**0.827**
	Graduation	54	60.93	13.118			
	Post Graduate/ Post Professional-MBA, CA, Doctor etc	45	61.67	13.555			
	Total	115	61.54	13.244			
Work Life Balance	<=12th Std.	16	47.13	8.884	2	0.54	**0.584**
	Graduation	54	49.7	9.009			
	Post Graduate/ Post Professional-MBA, CA, Doctor etc	45	48.38	10.188			
	Total	115	48.83	9.435			

(The mean difference is significant at the 0.05 level.)

From the table, it can be seen that on comparing the average scores of Husband's educational qualification and Achievement Motivation of respondents, there is greater significant difference (p-value = 0.022). No significant difference is observed for leadership (p-value = 0.827) and work life balance (p-value = 0.584) when respondents average score is compared with Husband's Educational Qualification. Therefore, it can be interpreted that there is greater significance between Husband's Educational Qualification and Achievement Motivation of the respondents. However, no significant difference is seen between Leadership and Work life Balance of respondents and Husband's Educational Qualification.

To further analyze the correlation between the significance of variables, the data was exposed to Post-Hoc Test which used the Scheffe method to determine the comparisons and revealed the following results:

Table 6.10.1 Showing multiple comparisons between Husband's Educational Qualifications and Achievement Motivation

N=115

Scheffe				
Dependent Variable	(I) Husband's Education Qualification	(J) Husband's Education Qualification	Mean Difference (I-J)	Sig.
Achievement Motivation	Graduation	Post Graduate Professional-MBA, CA, Doctor etc	6.041*	0.023
*The mean difference is significant at the 0.05 level.				

It can be seen from the above table that Achievement Motivation is highly significant between the respondents whose husbands have an educational qualification as Graduation and those that have pursued education as Post Graduation/ Professional (MBA, CA, Doctor, etc.) (p-value=0.023).

Here, N=115 as 37 respondents are not married while 06 respondents chose not to respond to this question.

Table 6.11 Showing correlation between Number of Children of respondents and Achievement Motivation, Leadership and Work life Balance

N=158

		N	Mean	Std. Deviation	df	F	Sig.
Achievement Motivation	0	14	60.14	9.281	6	1.704	0.124
	1	51	63	11.216			
	2	50	59.42	10.772			
	3	7	58.14	10.107			
	4	3	54.67	14.012			
	NR	5	65.4	6.504			
	NA	28	65.43	7.686			
	Total	158	61.75	10.387			
Leadership	0	14	62.86	13.761	6	0.393	0.883
	1	51	62.98	12.588			
	2	50	59.72	14.555			
	3	7	59.29	8.015			
	4	3	57.33	28.378			
	NR	5	63.4	5.899			
	NA	28	61.57	10.031			

	Total	158	61.43	12.873			
Work Life Balance	0	14	43.64	8.473	6	1.331	0.247
	1	51	50.94	10.51			
	2	50	48.94	8.959			
	3	7	44	8.563			
	4	3	45	15.875			
	NR	5	48.8	10.803			
	NA	28	47.89	12.02			
	Total	158	48.63	10.261			

(The mean difference is significant at the 0.05 level.)

From the table, it can be seen that by comparing average scores of Number of Children of respondents on Achievement Motivation, Leadership and Work life Balance, there is no significant difference (p-value = 0.124, 0.883 and 0.247).

Therefore, it can be interpreted that there is no significance between Number of children of respondents and Achievement Motivation and Work life Balance.

Table 6.12 showing correlation between Average Family Income (Per Month) and Achievement Motivation, Leadership and Work life Balance

N=103

		N	Mean	Std. Deviation	df	F	Sig.
Achievement Motivation	<=Rs.10,000	3	58.33	0.577	3	0.477	**0.699**
	Rs.10,001-20,000	5	66.6	14.17			
	Rs.21001-50,000	23	63.39	6.207			
	> Rs. 50,000	72	63.04	10.225			

	Total	103	63.16	9.5			
Leadership	<=Rs.10,000	3	43.33	4.041	3	3.5	**0.018**
	Rs.10,001-20,000	5	54.6	11.502			
	Rs.21001-50,000	23	60.7	11.384			
	> Rs. 50,000	72	63.32	12.163			
	Total	103	61.73	12.284			
Work Life Balance	<=Rs.10,000	3	31.33	3.055	3	5.846	**0.001**
	Rs.10,001-20,000	5	44.6	12.462			
	Rs.21001-50,000	23	48.65	11.511			
	> Rs. 50,000	72	51.93	8.294			
	Total	103	50.24	9.869			

(The mean difference is significant at the 0.05 level.)

From the table, it can be seen that by comparing average scores of Average Family income on Achievement Motivation, there is no significant difference (p-value = 0.699) of average scores by Average Family Income. Greater significant difference is observed on comparing respondents average scores on Average Family Income and leadership (p-value = 0.018) and Average Family Income. Greater significant difference is seen on comparing average scores of respondents on Average Family Income and work life balance (p-value = 0.001) and Average Family Income. Therefore, it can be interpreted that there is greater significance between Average Family Income and Work life Balance of the respondents. However, no significant difference is seen between Leadership and Achievement Motivation of respondents and Average Family Income.

To further analyze the correlation between the significance of variables, the data was exposed to Post-Hoc Test which used the Scheffe method to determine the comparisons and revealed the following results:

Table 6.12.1 Showing multiple comparisons between Income of Family (Per Month) and Leadership and Work life Balance

N=103

Scheffe				
Dependent Variable	**(I) Income of Family (Per Month)**	**(J) Income of Family (Per Month)**	**Mean Difference (I-J)**	**Sig.**
Leadership	<=Rs. 10,000	> Rs. 50,000	-19.986	.048
Work Life Balance	<=10,000 Rs.	Rs. 21001-50,000	-17.319*	.030
		> Rs.50,000	-20.597*	.004
* The mean difference is significant at the 0.05 level.				

From the above table, it can be interpreted that respondents Leadership is significant between those respondents whose Average Family Income (Per Month) is Rs. 10,000 with those whose Average Family Income (Per Month) is Rs.50,000 (p-value=0.048). Work life balance is highly significant with those respondents whose Average Family Income (Per Month) is Rs. 10,000 or more with families whose Average Monthly income is Rs. 21,000-50,000 (p-value=0.30) and with those whose Income is Rs. 50,000 (p-value=0.004).

Here N=103 as 55 respondents were hesitant in disclosing their family income.

The above exposed data gives a very vivid picture of how the dependent variables are associated with the independent variables in the current study. Clarity in terms of the interrelatedness of variables is also got from the data set.

A strong conclusion which can be drawn from the quantitative data is that each entrepreneur has their own set of skills and struggles and it takes a great deal of courage to continue working despite any fruitful result visible. A lot of decisions taken while enterprising may not be immediately rewarding but they gradually enhance the capacity of the women to appreciate their own working style and their own actions.

Herein, the words of Caroline Fernandes, Co-founder, Vibha can be quoted wherein she talks about women entrepreneurs as, "To me, a woman who is financially independent is very powerful. She has the ability to say YES. She also has the ability and strength to say NO."

Based on the data obtained, the researcher has developed selected case studies, one from each zone explaining the differences in the approaches of respondents and also showing the different areas in which respondents choose to pursue entrepreneurship.

SECTION VI

Case Studies from Selected Respondents:

During the early months of 2018, the respondent was muddling with the data collection after making all possible efforts to contact the right kind of respondents, the researcher reached out to 158 respondents across the four cities of Ahmedabad, Rajkot Surat and Vadodara. The city-wise break up of respondents is as under:

S. No.	City	No. of Respondents
1.	Ahmedabad	50
2.	Rajkot	19
3.	Surat	50
4.	Vadodara	39
TOTAL		**158**

Each respondent had a unique experience to share and almost everyone had a journey full of struggles and setbacks. But what made them what they are today is their perseverance and patience and the undying will to never give up. The researcher has tried to incorporate few narratives from the respondents as part of the study to show how women work in different trades and enterprises. The researcher has taken care of the ethical aspects and considerations before producing the case studies. Owing to these ethical considerations, the names of the respondents have not been disclosed for the purpose of confidentiality. Thus, narratives have been framed to display the issues and challenges faced by the respondents. The researcher has also attempted at trying to correlate the observations and notes along with the Social Work Skills to identify how these skills have been used by the respondents, and to what extent. Despite the pool of data available, the researcher has limited to sharing one case study per zone to justify the collected data as well as to show variety in the approach of respondents within the same state.

Making the best out of waste

1. **Name of the Organization:** WastePro, Ahmedabad
2. **Year of Establishment:** 2011
3. **Objective:** Understand, Update, Upgrade and undertake all steps possible in biodegradable wat=ste treatment and management.
4. **Activities:** The WastePro team is a team of spirited individuals committed to change the way citizens view waste. To ensure their objectives are met, the team empower children and citizens to become change agents themselves. They also bring the Do It Yourself (DIY) concept in Waste Engagement. Moreover, by this, contribute significantly to Swachh Bharat Mission.
5. **Entrepreneurial Journey:** "I was fed up of the inaction of the municipal corporation & watching cows feeding on plastic waste in their search for food. That is when it first struck me to begin a movement to address this menace and turn into a social entrepreneur," says the respondent.

   The respondent founded her for-profit enterprise in the year 2011, after a career as an Employee for 9 years. During her employment tenure, she worked for credible organizations like ITC Ltd, Confederation of Indian Industry (CII), Autonomous University of Madrid, Spain, Mission Mangalam of the Commissionerate of Rural Development, Govt. of Gujarat and lastly the Gujarat Chamber of Commerce & Industry. The burning desire to turn an employer and start her company led the respondent to tread the uncharted path of a new venture creation specialising in Solid Waste Management. Rearing to venture a male-dominated and government-centric sector was not an easy decision for her as a woman entrepreneur. But, today as an official stakeholder, she has treated tonnes of waste and prevented waste.

   The business model of the respondent is that of treating waste in a decentralized manner, at the point of generation itself. As the project treats a unit's waste in their own premises, garden, food & paper waste is prevented from reaching landfill sites. The venture promotes decentralization of solid waste treatment and management. The enterprise deals with the sales of waste handling tools, vermicomposting tools, earthworms and bio culture.

6. **Beneficiaries**: The respondent has taken her ideas to generate revenues by selling Bioculture, Kitchen waste composting kits, selling earthworms & selling vermi-compost to bungalows, farmhouses, industries & organic farmers. Karnavati Dental School, Mayur Greens Farm House, Shri Uma Industries, Naroda and J K Lakshmi Cement Ltd., Kalol are few of the many beneficiaries.

7. **Issues and Challenges:** The respondent went home to home teaching residents to handle their waste. She has not only toiled to make Ahmedabad a greener city but she has also taught people to derive wealth from waste. She has enlightened many school going children on environment conservation and has transformed lives of many people.

   The biggest challenge that the respondent and her enterprise face today is shifting human consciousness. She believes that it is not the planet that needs saving, it is people. She believes if people start taking environment for granted, there will soon be a time when human race might just be wiped off from the face of the Earth.

8. **Correlation with Dependent Variables:** The respondent's future aspiration is to be known as one of the key social venture companies instrumental in laying the foundation for a Zero Waste Amdavad by 2030.

   The respondent works individually managing her finances and operations with quite a balance. She has a team of workers and labourers not co-workers. She finds it difficult to balance her work-life and personal life; hence, she chooses to live a life by integrating work and family thereby giving her a flexible social life.

9. **Social Work Skills Used (as perceived by the researcher)**: The researcher observed the skills of information gathering and analysis in the respondent. The researcher also observed that the respondent was assertive and direct in her approach. She considers herself as a one-person army and prefers working solo. She is good at problem solving. She has good skills of delegating the work and is inspirational to people to become more environment-conscious.

10. **Learnings:** The respondent prefers to be called a Social entrepreneur. She decided to quit her well-paying job and invest all her savings as seed money for her venture. She believes that nature does not need human beings rather human beings need nature. She has won the best stall award at Vibrant Gujarat 2013.

She was a part of the stakeholder's meeting in drafting the new solid waste management rules 2013 of the Ahmedabad Municipal Corporation.

Create your own destiny!

1. **Name of the Organization:** Krish Consultancy, Rajkot
2. **Year of Establishment:** 2006
3. **Objective:** To be the Most Preferred Training and Placement Provider at India Level.
4. **Activities:** The enterprise began with two services - Job Placement (Recruitment and Staffing Solutions) and .HR & Training Implementation Services for Industries and Corporate Houses.
5. **Entrepreneurial Journey:** After Graduating, the respondent decided to start Placement Consultancy firm as she observed that people are willing to work but they do not get the chance or do not have required skills and appropriate information. So in 2006, the respondent started her own consultancy which has now grown to 25 associates across the state, 100 clients and a data base of 1 lac candidates. "Women are generally perceived as home makers with little to do with economy or commerce. However, this picture is changing. In Modern India, more and more women are taking up entrepreneurial activity especially in medium and small scale enterprises", says the respondent. She provides customized Human Resource Management and Training solutions through project based arrangements or complete outsourcing of the human resource function. She has vast experience to identify capability & skill of individual to develop their job career. She is also associated with various Management Institutes and Schools as a Resource Person and Consultant.

   She is passionate to provide best job and career counselling facilities to every job seeker of Saurashtra region through her company. She has served more than 100 Industries & corporate as a HR & Training Consultant. She is a youth icon for youngsters of Saurashtra region, especially the females.
6. **Beneficiaries**: The consultancy has grown over the years and it has been pivotal in providing jobs to over 1 lac candidates within the Saurashtra region. She has provided trainings in almost 100 industries and corporate houses. She has trained more than 10,000 Executives/supervisors in different industries & more than 50,000 students of Institutions.
7. **Issues and Challenges:** She feels that in the region where she belongs, the parents' goal is to educate the females to ensure they get the right match. This

mind set hinders many women who have entrepreneurial ideas. Thus, the region sees less contribution of women in entrepreneurial activities as compared to cities like Ahmedabad and Surat.

8. **Correlation with Dependent Variables:** The respondent has contributed in growth & development of SMEs in last 13 years as 5S Expert. She is self-motivated and found space for her venture because there was a need for such establishment in the market. She is a corporate trainer and leads others be example. She has a dedicated team of trainers and placement providers who work in cohesion to provide employment opportunities to the unemployed youth. Her enterprise follows the model of providing skill-based jobs. Since the enterprise has been long establishes, the respondent finds it very easy to strike balance between her work life and her family life. .
9. **Social Work Skills Used (as perceived by the researcher)**: The researcher observed that the respondent possessed excellent skills of communication and presentation. She possesses right knowledge and owes to her team to keep her updated on the latest news and trends that help her in her training sessions. She is resourceful and has a well-knit network of professionals who push her to fame. She possesses great leadership skills and treats her workers as family. She is empathetic and assertive. She does not believe in jeopardising personal relations at the cost if professional negligence and hence takes utmost care in dealing with her employees. She possess good interpersonal skills and problem-solving skills. She maintains a calm and pleasant disposition.
10. **Learnings:** The respondent believes that life is all about creating yourself so one must design his or her own destiny. The researcher learned that it takes years of struggle to be able to create a space for oneself in the market. The researcher observed this undying spirit to succeed at all costs, in the respondent.

Inspiring people to live WAKEFULLY!

1. **Name of the Organization:** Happy Minds, Surat
2. **Year of Establishment: 2012**
3. **Objective:** The sole mission and objective of the company is to transform lives.
4. **Activities:** HappyMinds is a multi – faceted organization that is committed to bringing this change by empowering people coming from different walks of life. HappyMinds deals in following three verticals:

   ProMinds: These Training and Workshops aim at mentoring & training individuals & organizations to enable them to reach a higher level of thinking thereby giving a meaning to their existence.

   Leadership Academy: The Leadership Course will take you from existence to excellence! To create an impact and change the approach from traditional bookish learning to global age practical learning.

   Social Initiative – Jazbaa: This vertical is driven by the Jazbaa to make a difference to society in unique ways. Sessions on iLearn to iLead, Nurturing Young Trainers, Positivity Drive, etc. are the core fundamentals of this vertical.
5. **Entrepreneurial Journey:** The respondent pursued her Master's Degree in Business Administration from MIT, Pune. There she chose a module which, allowed her to work while studying and got herself placed in a Corporate training company. Her work profile was in the marketing sector but she stretched her working hours to learn as much as possible. She spent her break-times in seminars, where one of her senior used to deliver sessions on 'Life Skills'. She happily served tea and snacks to the participants just to be present there and simultaneously, learn. Even after, tiring working hours, she used to manage her time and make modules from what she learnt in the whole day. Since the beginning, she has followed the mantra – "Wake up, Dress up and Show up". And keeping the same mantra as her motivation, she soon started off as a trainer; however, small the number of audience, she had her mother and her best friend, as her constant. There were times, when people tried to pull her down. They mocked her as some 'modern-day-baba' and went on to call her a 'fanatic'. But she had a clear positive vision and a huge faith in her own work. She had built a strong inner system, relying on which she kept going on and did not let criticism disturb her pace. She then studied NLP (Neuro-Linguistic

Program) where the roots of her ideology were nurtured. She then invested herself completely in understanding the thought patterns of the human mind.

6. **Beneficiaries:** HappyMinds offers services like training, seminars and workshop sessions and events on positive parenting, entrepreneurial empowerment, confidence building, communication, homemakers' empowerment, corporate training, youth conferences, positive thinking sessions and many more. The beneficiaries range from the ages of 4+ to 60+ years.
7. **Issues and Challenges:** From going door to door to share her ideas, to giving free seminars and training, the respondent dedicated her initial entrepreneurial years in spreading awareness on mind power and building a strong base of trust and understanding among her growing audience. After almost, half-a-decade of persistent efforts and hard work, she brought the wave of change she always intended to. Now, when she announces her public seminars on either 'Positive Thinking', 'Happiness Therapy', 'Mindpower' or any other Life skill subjects. The respondent believes that even today, there is not much awareness on the importance of mental wellbeing. She believes a lot needs to be done and she needs to expand the horizons of her venture to different parts of the country and the world at large.
8. **Correlation with Dependent Variables:** The respondent at the age of 19, dreamt and planned out a certain kind of life for herself and did every possible thing to make it work. However, fate had other plans altogether, none of it worked out the way she wanted. Broken and shattered, she started once again. This time her mind was railing with the idea of understanding how it all works. She intended to know how the mind over matter changed the game. That is precisely the time when things started playing out in her favour. This shows the level of motivation of the respondent. She is highly motivated by her own success and aspires to do more to make the world a better place. From a trainer, she has now become an influencer.

   The organisation has a flat model, having more than 25 volunteers associated with them. The respondent, who is the boss, chooses to be called Director and Chief Mentor while she prefers to call her staff members as Chief Energy Officers (CEOs). She takes fun seriously and ensures all employees are always happy and cheerful. She follows a leadership model that further empowers the staff and volunteers to stay committed to the cause of transformation.

At an early age of 24 years, the respondent was the youngest speaker on the panel across the country for her Social Contribution at the TEDx TALK platform.

BRITT India and iHeart Global Inspiration Women'17 have honoured her as one of the top 60 women entrepreneurs in India with innovative business ideas in 2016.

The respondent owes her success to her extremely supportive mother, her loving husband and both her families. Moreover, the positivity that her professional life gives her makes her personal life bloom too. Hence, she does not face any significant work-life balance issues.

9. **Social Work Skills Used (as perceived by the researcher)**: The respondent is an extremely confident individual with a very clear understanding of herself and her work. She lives a purpose-driven life and chooses to transform people's lives through her efforts. From the perspective of social work skills, the researcher has observed that the respondent possess excellent skills of communication and presentation. She has great resource mobilization skills and very sound interpersonal as well as intrapersonal skills. She follows the POSDCORB model thoroughly and is an efficient manager. She possesses good problem solving skills. She is accepting of herself and the people around her and believes in holding a non-judgmental attitude towards others. She is an epitome of positivity, enthusiasm and energy.

10. **Learnings:** The researcher learnt about a new area of entrepreneurship, which dealt with infusing leadership among children as young as 4 years. On her visit to the enterprise, the researcher observed how intricately confidence boosting and leadership training is provided to the young minds. The researcher also observed how patiently and calmly the staff engaged with children in their fun activities. It was surprising to see that many parents are actually interested in the leadership camps for children and are willing to pay any amount just to enrol into the HappyMinds Academy.

Serving bowlful of happiness

1. **Name of the Organization:** Messy Bowl (earlier Maggewala), Vadodara
2. **Year of Establishment:** 2013
3. **Objective:** To pioneer in delivering hand-crafted food to people and savor their taste buds.
4. **Activities:** The USP is to create usual instant noodles with a twist and offer variety of flavours for all age groups. Their dishes are served at competitive and affordable prices. Core reason of our good standing in the market is our quality that has authentic flavours. It helps them become honest & trust-worthy brand among Maggi-lovers.
5. **Entrepreneurial Journey:** The respondent ventured into food industry without any prior experience. The respondent was working at the Vadodara Municipal Corporation making a sound living with a regular paycheque rolling in every month. Her life was near to perfect. She was also married to the person of her choice, which made her feel that her life was complete. However, life had a different take on her perceived happiness. She soon realised she was in to a troubled marriage and wanted to help her husband by running her venture. However, her husband belonged to a family conditioned to believe that a stable job is what one needs and abhorred the aspect of entrepreneurship. Moreover, the husband was not happy in letting his wife work while he had no income. However, the respondent stood against all odds to make sure her venture succeeded. Her family and friends immensely supported her and she truly believes that her legal separation from her life partner was like receiving a new life. She is an example of what a woman, who has only self-belief by her side, can do. With her vision to serve best in class quality food at affordable prices, she started a brand, that is today, loved by people of all age groups.
6. **Beneficiaries:** From one outlet in the heart of the cultural capital of Gujarat (Vadodara), it became one of the fastest growing chain in the fast-food industry. She has created a pool of happy customers in various parts of the country. Besides having their outlets in Vadodara, the venture has also expanded into cities of Rajkot, Surat, Hyderabad and Chennai.
7. **Issues and Challenges:** Founded in 2013 by a first generation entrepreneur with a belief in pioneering in delivering handcrafted food to people and savour

the taste buds. Core reason of their strong foothold in the market, despite selling varieties of only one product, is their quality that has authentic flavours. This strategy helps them become honest & trust-worthy brand among Maggi-lovers. "We do not have a fancy, elaborate menu but whatever we serve is cooked with all our heart", says the respondent.

They faced a big challenge when, in 2015, the Nestle Group landed under a legal scanner being accused of being unsafe and hazardous for consumption. As against the maximum limit of 2.50 parts per million (ppm), the amount of lead detected in the Maggi samples was perilously high at 17.2 ppm. The second offence was mentioning 'No added MSG' on the packaging, which is an act of mislabelling. On June 5, 2015, a nationwide ban was imposed on Maggi by FSSAI (Food Safety and Standards Authority of India). During this phase, the respondent had to go out of business for a while s there was a public outrage on serving a banned product. The respondent had to change the menu at her restaurant chain and had to wait until the matter cleared. That was a time when the respondent's outlet was named Maggewala. Once Nestle re-established itself back in the market, the respondent changed the name of her outlet from Maggewala to Messy Bowl. She continues to serve the same product with guaranteed safety. The product and the food chain gained such popularity that she was invited as a guest at the TV Show - Maggi Kitchen Journeys, hosted by the renowned Bollywood actress Smt. Renuka Shahane Rana (https://www.youtube.com/watch?v=JDIKH2Ffx6I).

8. **Correlation with Dependent Variables:** The respondent kept a very low profile and was happy with her life. After her marriage and shift to Ahmedabad, she came across many street vendors who were experimenting and serving Maggi noodles. On one such evening, when she was out with her friends sharing a bowl of Maggi, it occurred to her to leverage on this idea and open a chain in her hometown – Vadodara. Things seemed to be falling in place, until she found out that her husband did not have a job and she stepped forward to help with her venture. She was shocked to learn that the husband was not in favour of her working and restricted her movement. It was then that her mother stepped forward and helped her come out of a failed marriage. Leaving this behind, the respondent felt more confident in taking the leap and was motivated to devote more hours to her venture.

The respondent was supported immensely by her family and few friends who helped her choose the right location, staff, furniture, marketing, advertising, publicity and streamlining things for her. She believes in bringing people together and leads by example. She feels if one provides proper attention to the details, it can win many hearts. She does not believe in authority and treats her staff as family.

Since initially the venture required more involvement, the respondent had to spend days and nights at the restaurant ensuring that everything was in order. At that time, it was difficult to manage her work life and her family life. Gradually, when things started working out favourably, she was able to take time off for her personal commitments. Today, she has a well-balanced life at both personal and professional fronts.

9. **Social Work Skills Used (as perceived by the researcher)**: The researcher observed that the respondent possessed great managerial skills. She had a vision and was able to persevere through all the troubles to make sure her vision turned to reality. She applies the POSDCORB model very efficiently. She displays excellent interpersonal and intrapersonal skills. She possesses sound leadership skills and good skills of resource mobilization. She thrives on her networking skills, which have led to the nationwide expansion of her outlets. She has great problem solving skills. She maintains a pleasant disposition and is always ready to continue to inspire young and potential entrepreneurs with her self-belief and positivity.
10. **Learnings:** The researcher learned that eureka moment could also occur during the times one least expects it. In case of the respondent, it occurred over a bowl of Maggi. Today, that has turned into a successful business venture. The researcher learned that entrepreneurial success does not come overnight and entrepreneurs have to persevere hard to ensure that their vision becomes a reality. The researcher also learnt that despite the difficulties, one needs to pursue and put their foot down to create great results.

Concurring from this data, the researcher provides findings and suggestions in the next chapter.

CHAPTER V

FINDINGS,

DISCUSSION

&

CONCLUSION

FINDINGS, DISCUSSION AND CONCLUSION

In this chapter, the researcher has attempted to state the major findings of the study which gradually culminate towards the acceptance or rejection of the hypotheses assuming which the current study was undertaken. The chapter is divided in to two parts -

Part A

This part is further divided into four sections:

Section I - Major Findings from the Primary Information of Respondents

Section II - Findings from the Variables of the Current Study

Section III - Findings from Interpretation of *f*-Test {Analysis of Variance (ANOVA)}

Section IV - Acceptance or Rejection of Hypothesis

Part B

This part focuses on the discussion based on the obtained data and ends on concluding remarks given by the researcher on the basis of findings and other experiences during the journey of culminating the doctoral work.

This chapter gradually paves its way into the forthcoming chapter which reads the suggestions and recommendations offered by the researcher in form of a module associating the Methods of Social Work and the current study.

Part A

Section I

Findings from Primary Information of Respondents

The current study followed certain objectives in order to scientifically assess every aspect under study. The first objective of the study was to study the socio-economic background of women entrepreneurs in Gujarat. In order to gather the socio-economic data of the respondents, the researcher had questions incorporated within the tool of data collection which would help bring out the correct status of women entrepreneurs across the four cities of Gujarat - Ahmedabad, Rajkot, Surat and Vadodara.

From the total respondents, the background information such as Age, Education, Marital Status, Average Monthly Income of the respondent, Average Monthly Income of Family, Husband's Education, Husband's Occupation, Number of Family Members, Number of Children etc. was calculated which showed the following results.

- 59 (37.3%) respondents fall between the age groups of 31-40 years.
- 71 (44.9%) respondents have pursued education up to Graduation level whereas 52 (32.9%) of the respondents have pursued education up to Post Graduate level.
- Majority of the respondents i.e. 77.2% are married.
- 54 (34.2%) respondents have an experience of more than 10 years in running an enterprise.
- 60 respondents, i.e., 38% have an experience of 2-5 years in running their current enterprise whereas 38 (24.1%) respondents have an experience of running their own enterprise for the last 6-10 years.
- 41 respondents, i.e., 25.9% have an average monthly income of Rs. 21,001 - 50,000. A sizeable number of respondents, 32 (20.3%) have answered that their monthly income is not regular.
- 43 respondents' husbands have their own business i.e. 27.2% and an equal per cent belong to professional category. Professions like Chartered Accountants, Doctors etc. are considered a part of the Professional category.

- ❖ While 39 respondents, i.e., 25.3% have chosen not to disclose their husband's average monthly income, 24.7% respondents have claimed that the average monthly income of their husbands is more than Rs. 50,000/-.
- ❖ 57 respondents, i.e., 36% have husbands whose minimum qualification is graduation.
- ❖ Majority of the respondents, i.e., 81.6% have 3-5 members in their families.
- ❖ 51 respondents, i.e., 32.3% have only one child.

It was important to collect primary information as these factors have an impact on the entrepreneurial abilities of the respondents. The focus of the study undertaken was to find factors which increase the Achievement Motivation of the respondents, factors which bring out the Leadership skills and those factors which help in understanding the Work life balance of the respondents.

The information provided by all the respondents turned out to be very useful in determining the factors under study.

Section II

Findings from the Variables under study

Apart from the socio-economic background of the respondents, the researcher had also designed the tool for data collection which could help in analysing the motivational factors behind women entrepreneurship, identifying the leadership quality among women entrepreneurs in Gujarat and find the degree of work life balance of women entrepreneurs in Gujarat.

After exposing the data to various statistical packages and obtaining analysed results, the following findings, according to the objectives are presented in this section.

2. To analyse motivational factors behind women entrepreneurship.

The Achievement Motivation tool consisted of 17 items. Each respondent's response was collated and cumulated in to one table. (Chapter III - Tables 2.1 and 2.1.1). The findings are as below:

- Out if the total number of respondents, majority of the respondents, i.e., 57% agree to some extent on having Achievement Motivation, a lesser frequency of 33.5% mostly agree that they have high Achievement Motivation and only about 9.5% of respondents disagree on having Achievement Motivation.

Since Achievement Motivation is one of the dependent variables of the study in concern, this factor was further correlated with each of the independent variables of the study to generate the following results:

- The correlation between Age, an Independent Variable and Achievement Motivation, a Dependent Variable gives the Chi-Square Value as 21.578[a] and significance level 0.001. This means as respondents age increases, their motivation to work on their business and also increases.
- The correlation between Education, an Independent Variable and Achievement Motivation, a Dependent Variable gives the Chi-Square value as 13.630[a] and significance level 0.092. It does not

show any significant correlation. This clearly emphasizes that Achievement Motivation does not significantly depend on Education.

- ❖ The correlation between Experience, an Independent Variable and Achievement Motivation, a Dependent Variable gives the Chi-Square Value for association as 20.806[a] and the Significant p-Value as 0.002. This reveals that the above variables are highly correlated with each other.

This signifies that women entrepreneurs tend to gain more Achievement Motivation with increased experience in the field.

- ❖ The correlation between Average Income (Per Month), an Independent Variable and Achievement Motivation, a Dependent Variable shows the Chi-Square Value for association as 20.699[a] and the Significant p-Value as 0.023. It can be inferred from this data set that higher income develops greater achievement motivation among women entrepreneurs.

3. To identify the leadership quality among women entrepreneurs in Gujarat.

The Leadership Scale consisted of 18 items. Each respondent's response was collated and cumulated in to one table (Chapter III - Tables 3.1 and 3.1.1). The findings of respondents on Leadership are as follows:

- ❖ Out of the 158 respondents, majority of the respondents, 75.3% have expressed greater leadership abilities while only about 24.7% of respondents expressed that they have experienced their own leadership abilities sometimes.

Although the tool followed three pointer-scale and the respondents were asked to respond within the categories of 'Never', 'Sometimes' and 'Always' but the cumulative responses showed that the respondents did not respond in negation, hence only the other two categories were considered.

Since Leadership is one of the dependent variables of the current study, this factor was further correlated with each of the independent variables of the study to generate the following results

- ❖ The correlation between Age, an Independent Variable and Leadership, a Dependent Variable reveals the Chi-Square Value as 1.609[a] and Significant p-Value as 0.657. The variables are not significantly correlated with each other. Hence, Age is not correlated with Leadership.
- ❖ The correlation between Education, an Independent Variable and Leadership, a Dependent Variable reveals the Chi-square Value as 5.695[a] and the significance as 0.223. The variables are not significantly correlated with each other. Hence, Education is not correlated with Leadership.
- ❖ The correlation between Experience, an Independent Variable and Leadership, a Dependent Variable reveals the Chi-Square Value as 8.627[a] and Significant p-Value is 0.035. Since the obtained score is 0.035 which is lesser than the accepted mean difference, the two variables are accepted as significantly correlated with each other.
- ❖ From the obtained data for the current study, The hypothesis (H_2) that - With increased Experience, Women Entrepreneurs develop their own Leadership is accepted within the purview of this study.
- ❖ The correlation between Average Income (Per Month), an Independent Variable and Leadership, a Dependent Variable reveals the Chi-Square Value as 10.950[a] and Significant p-Value is 0.052. Since the mean difference is significantly accepted at 0.05 level and the obtained score is 0.052, the variables are partially significant and correlated with each other.
- ❖ It can be inferred from the obtained data that with an increase in the Income, the leadership is also enhanced. Hence, Income and Leadership are partially significant with each other.

4. **To find the degree of work-life balance of women entrepreneurs face.**

The tool designed to find out the work life balance of respondents had 20 items with a five-pointer scale response such as - 'Not true', 'Little True', 'Somewhat True', 'Fairly True' and 'Definitely True'. The total responses of the respondents were cumulated into one table (Chapter III - Tables 4.1 and 4.1.1).

- It is Fairly true for majority of respondents, 61.4% to achieve work life balance whereas for 24.7% respondents, achieving work life balance is somewhat true, only a small segment of respondents, 13.3% respondents are able to definitely strike a good balance between their work life and their family life whereas for only a small population of 0.6%, it is difficult to achieve work life balance.

As one of the dependent variables chosen for this study is also work life balance, it was further correlated with the other independent variables to find out the correlation. By exposing data sets to various statistical tests, the following results were obtained.

- A correlation between Age, an Independent Variable and Work life Balance, a Dependent Variable reveals the Chi-Square Value as 13.367[a] and Significant p-Value is 0.147. Since the mean difference is significantly accepted at 0.05 level and the obtained score is 0.147, the variables are not significantly correlated with each other. It can be inferred from the data obtained that Work life balance of Women Entrepreneurs has no correlation with their Age. Hence, Age and Work Life Balance are not significantly correlated with each other.
- A correlation between Education, an Independent Variable and Work life Balance, a Dependent Variable brings out the Chi-Square Value as 10.316[a] and Significant p-Value is 0.558. Since the mean difference is significantly accepted at 0.05 level and the obtained score is 0.558, the variables are not significantly correlated with each other. It can be inferred from the obtained data that Work life balance of Women Entrepreneurs has no

correlation with their Education. Hence, Education and Work Life Balance are not significantly correlated with each other.

- A correlation between Experience, an Independent Variable and Work life Balance, a Dependent Variable shows the Chi-Square Value as 15.587[a] and Significant p-Value as 0.076. Since the mean difference is significantly accepted at 0.05 level and the obtained score is 0.076, the variables are not significantly correlated with each other. It can be interpreted from this data that Work life balance of Women Entrepreneurs has no correlation with their Experience in the field. Hence, Experience and Work Life Balance are not significantly correlated with each other.
- A correlation between Average Income (Per Month), an Independent Variable and Work life Balance, a Dependent Variable reveals the Chi-Square Value as 114.149[a] and Significant p-Value is 0.000. Since the mean difference is significantly accepted at 0.05 level and the obtained score is 0.000, the variables are highly significant and correlated with each other. It can be inferred from the obtained data that Work life balance of Women Entrepreneurs has significant correlation with their Income.
- From the obtained data for the current study, the hypothesis formulated that - **Work life Balance is absent among Women Entrepreneurs (H_3)**, is not accepted. This concludes that despite other factors not letting respondents achieve a balance between their work life and their family life, higher income helps them to do so. Hence, Income and Work Life Balance are significantly correlated with each other.

Section III

Findings from Interpretation of *f*-Test {Analysis of Variance (ANOVA)}

In the previous section, the researcher has correlated the independent and dependent variable by exposing them to Chi-Square Tests and obtained the significance values of the dependent and independent variables.

However, the data set was also further exposed to f-Test for finding out the Analysis of Variance among data and the generated results have been explained in this section.

The following are the results explained on f-Test:

- On comparing average scores by Age groups, it is seen that the Achievement Motivation and Work Life Balance of respondents are significant (p-values=.004 and 0.048) where as it is not significant for the Leadership. It can be inferred that there is greater significance between Age and Achievement Motivation and partial significance between Age and Work life Balance but no significance is seen between Age and Leadership. This brings us to an understanding that both Achievement Motivation and Work life Balance tend to increase with age, among the respondents. On further analysis of the significant values through the Scheffe method of Post-Hoc, it was found that that there is greater significance of Achievement Motivation within the age group of 31-40 years (p-value=0.024) in multiple comparison with respondents of 51 years and above.
- There is no significance between Educational Qualification and Achievement Motivation, Leadership and Work life Balance of the respondents (p-values=0.130, p-value=0.216 and p-value=0.469 respectively). It can be interpreted that there is no significance between Educational Qualification and Achievement Motivation, Leadership and Work life Balance of the respondents.
- There is very little significance between Marital Status and Achievement Motivation (p-value=.056) but no significant difference due to respondents' marital status in Leadership (p-value=0.599) and Work life Balance. (p-value=0.737).

- There is no greater significance between Number of Years of Experience and Achievement Motivation, Leadership and Work life Balance of the respondents (p-values=0.301, 0.630 and 0.607 respectively).
- There is greater significance between Number of Years of Experience of running own enterprise and Achievement Motivation, Leadership and Work life Balance (p-values=0.001, 0.002 and 0.014 respectively). It can be interpreted that respondents have exhibited greater skills of leadership in running their own enterprise; they have also managed to balance their work life and family life and also maintained greater work life balance.

By applying Post-Hoc Studies using the Scheffe's method it is found that Achievement Motivation is highly significant with respondents having 6-10 years of experience with respondents having 2-5 years of experience (p-value=0.029) as well as those having more than 10 years of experience (p-value=0.002). Similarly, Leadership is highly significant with respondents having 6-10 years of experience with respondents having 2-5 years of experience (p-value=0.048) as well as those having more than 10 years of experience (p-value=0.005). This explains that leadership styles are developed by women entrepreneurs over time. And Work life Balance is highly significant with respondents having 6-10 years of experience with respondents having more than 10 years of experience (p-value=0.026) thereby signifying that with more number of years in their own business, women adapt to achieving more Work life Balance.

- On comparing average scores of Average Monthly Income on Achievement Motivation there is no greater significant difference (p-value = 0.744). On comparing scores of Average Monthly Income and Leadership, greater significant difference is observed (p-value = 0.007) whereas on comparison of average scores of Average Monthly Income and Work life Balance, greater significant difference (p-value = 0.000). Therefore, it can be interpreted that there is greater significance between Average Monthly Income of respondents and Leadership and Work life Balance. However, no significant difference is seen between Achievement Motivation and Average Monthly Income of respondents.

By applying Post-Hoc Test on the correlated variables, it was found that Leadership is highly significant with respondents having Average Income (Per Month) Rs <=10,000 with respondents having Average Income (Per Month) between Rs. 21,000 - 50,000 (p-value=0.058) as well as those having Average Income (Per Month) more than Rs. 50,000 (p-value=0.017).

Similarly, Work life Balance is highly significant with respondents having Average Income (Per Month) between Rs. 21,000 - 50,000 (p-value=0.000) and those having Average Income (Per Month) more than Rs. 50,000 (p-value=0.002). Also, significance is found between respondents having Average Income (Per Month) between Rs. 10,001 - 20,000 and Rs. 21,001 - Rs. 50,000 (p-value=0.005).

- On comparing average scores of Husband's occupation and Achievement Motivation there is greater significant difference (p-value = 0.039). However, no significant difference is seen for Leadership and Work life balance (p-value = 0.132 and 0.917) and Husband's Occupation. Therefore, it can be interpreted that there is greater significance between Husband's Occupation of respondents and Achievement Motivation. However, no significant difference is seen between Leadership and Work life Balance and Husband's Occupation of respondents.

By applying Post-Hoc Test using the Scheffe's method it is found that Achievement Motivation is highly significant with respondents' husbands Occupation as Job with those respondents whose husbands are also doing business (p-value=0.045).

- By comparing average scores of Husband's average monthly income on Achievement Motivation there is no significant difference (p-value = 0.726). No significant difference is seen on comparing the average scores of Husband's Average Monthly income and Leadership (p-value = 0.385). Greater significant difference is observed on comparing average scores for Work life Balance (p-value = 0.008) and Husband's Average Monthly Income. Therefore, it can be interpreted that there is greater significance between Husband's Average Monthly Income and Work life Balance. However, no significant difference is seen between Achievement Motivation and Leadership, and Husband's Average Monthly Income.

By applying Post-Hoc Test on the significant variables, it can be seen that there is there is partial significance within the variables of Husband's Income (Per Month) is Rs. 10,001 - 20,000 and between Rs. 21,001 - 50,000 (p-value=0.094) within the dependent variable of Work life Balance.

- ❖ There is greater significance between Number of Family Members and Leadership (p-value = 0.015). However, no significant difference is seen between Achievement Motivation and Work life Balance and Number of Family Members.

By applying Post-Hoc Test on the correlated variables, it can be seen that Leadership is highly significant between those respondents who have 1-2 family members with those who have 3-5 family members (p-value=0.036).

- ❖ On comparing the average scores of Husband's educational qualification and Achievement Motivation of respondents, there is greater significant difference (p-value = 0.022). No significant difference is observed for leadership and work life balance (p-value = 0.827 and 0.584) when respondents average score is compared with Husband's Educational Qualification. Therefore, it can be interpreted that there is greater significance between Husband's Educational Qualification and Achievement Motivation of the respondents. However, no significant difference is seen between Leadership and Work life Balance of respondents and Husband's Educational Qualification.

By applying Post Hoc Test on the correlating variables, it was found that Achievement Motivation is highly significant between the respondents whose husbands have an educational qualification as Graduation and those that have pursued education as Post Graduation/ Professional (MBA, CA, Doctor, etc.) (p-value=0.023).

- ❖ By comparing average scores of Average Family income on Achievement Motivation, there is no significant difference (p-value = 0.699) of average scores by Average Family Income. Greater significant difference is observed on comparing respondents average scores on Average Family Income and leadership (p-value = 0.018) and Average Family Income. Greater significant difference is seen on comparing average scores of respondents on Average Family Income and work life balance (p-value = 0.001) and Average Family Income. Therefore, it can be interpreted that there is greater significance between Average Family Income with Leadership and Work life Balance of the respondents. However, no significant difference is seen between Achievement Motivation of respondents and Average Family Income.

By exposing the data to Post Hoc Test for further correlation of significant variables, it can be analyzed that Leadership is significant between those respondents whose Average Family Income (Per Month) is Rs. 10,000 with those whose Average Family Income (Per Month) is Rs.50,000 (p-value=0.048). Work life balance is highly significant with those respondents whose Average Family Income (Per Month) is Rs. 10,000 or more with families whose Average Monthly income is Rs. 21,000-50,000 (p-value=0.030) and with those whose Income is Rs. 50,000 (p-value=0.004).

Section IV

Findings of Acceptance or Rejection of Hypothesis

The researcher began this research with certain presumptions and pre-conceived notions about women entrepreneurship. These presumptions were then devised into hypotheses for the current study. A hypothesis, in research, is a tentative statement designed by the researcher to provide insight in to the subject under study and provide direction to the study.

For the current study, the researcher had formulated the following hypotheses:

1. Higher the income, higher is the motivation to work (H_1).
2. With increased experience in the field, entrepreneurs develop their own leadership (H_2).
3. Work-life balance is absent in women entrepreneurs (H_3).

By exposing the data to various statistical tests and obtaining the results on various parameters, it was established that from the set hypotheses for the current study, two hypotheses were accepted as null hypotheses while one hypothesis was rejected.

Table 6.4.1 Showing Acceptance/ Rejection of Hypotheses

S.No	Hypothesis Statement	Accepted/ Rejected
1.	**Higher the income, higher is the motivation to work (H_1).**	**Accepted**
2.	**With increased experience in the field, entrepreneurs develop their own leadership (H_2).**	**Accepted**
3.	**Work-life balance is absent in women entrepreneurs (H_3).**	**Rejected**

It has been proven from the obtained data of the present study that income really plays a greater role in motivating the women entrepreneurs into working harder towards their business thereby helping them succeed. The result from the Chi-Square Value for association is 20.699[a] and the Significant p-Value obtained is 0.023. **Hence H_1 is accepted.**

It has been proven from the data that leadership is highly significant with experience of the respondents of running their own enterprises within the purview of this study. The correlation between Experience, an Independent Variable and Leadership, a Dependent Variable reveals the Chi-Square Value as 8.627[a] and the obtained Significant p-Value as 0.035. Leadership is also dependent upon Income, Occupation of the husband as well as partially with Number of Family Members of the respondents. There is greater significance between Number of Years of Experience of running own enterprise and Achievement Motivation, Leadership and Work life Balance (p-values=0.001, 0.002 and 0.014 respectively)

By applying Post-Hoc Studies using the Scheffe's method it is found that Achievement Motivation is highly significant with respondents having 6-10 years of experience with respondents having 2-5 years of experience (p-value=0.029) as well as those having more than 10 years of experience (p-value=0.002). **Therefore it can be established that H_2 is accepted.**

From the current study, it can be inferred that the women entrepreneurs have a balanced work life and family life. On correlating Work life Balance with the Independent variables, it was found that there is significant association between Age, Experience and Income and Work life Balance (p-values = 0.013, 0.035 and 0.000 respectively). Even on f-Test, Work life balance is significant with experience (p-value = 0.014) thereby signifying that with more number of years in their own business, women adapt to achieving more Work life Balance; it is also significantly correlated with income, whether it is their own or that of their husband and/ or of the family (p-values = 0.000, 0.008 and 0.001 respectively). From these significant findings it can be concluded that women entrepreneurs do strike a balance between their work life and their family life and that domestic responsibilities do not hamper their entrepreneurial performance and vice versa. **Therefore, H_3 is not accepted.**

Part B

Discussion and Conclusion

"You don't need a fancy degree to dream big and make it happen. It's all in your head, your heart your hands."

— Rashmi Bansal

With a thorough understanding of the above quote by one of the leading Indian authors on entrepreneurship, and that too a female, the researcher had embarked upon the journey of formulating a research on this small but highly effective population of women entrepreneurs. With an increased reading in this area and understanding the perspectives of people associated with entrepreneurship development, a greater insight and intrigue was developed to explore the psycho-social factors affecting women who are into entrepreneurship.

When our beloved and respected Prime Minister, Shri Narendrabhai Modi announced in his Independence day speech in 2015 about the Startup India Initiative which was then effectively launched in the year 2016, the researcher decided to plunge into this vast arena of entrepreneurship development especially for women. Another motivating factor which came across was the recent Ease of Doing Business Rankings 2019 wherein India was ranked 77th among 190 countries.

Having set the tone for understanding gender roles and responsibilities, the researcher chose to accept the challenge of identifying and studying women entrepreneurs of the state of Gujarat, setting a precedent that entrepreneurship is not just a male arena but females can also successfully handle it.

Gujarat is a land of opportunities. It is geographically blessed and hence is able to cater to both agriculture and industrialization to a large extent. Having the advantage of the longest coast also gives it an edge in terms of development of ports than other states of the country. Culturally speaking, Gujarat caters to a lot of diversity. This diversity has also helped bring about a change in the lives of the local people including the women. Women in Gujarat may be neither submissive nor rigidly aggressive. They are assertive and know how to get their work done. People of Gujarat are renowned for their sound business acumen that is evident in the females as well. This became one of the premises to choose the sample for the current study.

The researcher divided the sample from across the four cities of Gujarat - Ahmedabad, Rajkot, Surat and Vadodara. Each of these cities have an existence of a functional Chamber of Commerce and Industries and have association to which women entrepreneurs are associated. Apart from these agencies, the various Centers for Entrepreneurship Development located in the cities under study were also approached. However, the researcher was unable to find a breakthrough and hence decided to follow snowball sampling as a method of data collection. The researcher could easily lay hands on certain material regarding the directory of industries but there was no specific material available to know of exclusive women entrepreneurs. This material was only found in Ahmedabad while in Surat they did have a Ladies Wing but no directory; in Rajkot no such details were available and in Vadodara, making an exclusive directory for registered women entrepreneurs was in process. However, on individually contacting women entrepreneurs through reference, it was found that through a channel of internal networking, many women knew each other but there was no record of such data available.

Further on within the study, the researcher found a that an entrepreneur is truly someone who solves the same problems in a different manner and comes up with different solutions. The researcher has met many women in course of her doctoral work who have been the first generation entrepreneurs in their fields. While most women have enterprises related to more of feminine aspects, according to the societal norms, few women have taken a new road in establishing themselves in a market which was not dominated by them. Despite these oddities, they have succeeded in their ventures.

The husbands of these women have been supportive and encouraging. None of the respondents said to have begun enterprising without family support. There were very few cases of separation and some respondents were not married at the time of interview. But overall, the husband and his family members had always been supportive and considerate of the work done by them to economically support their families. This also brings to understanding that modern Indian families are now transitioning and have begun to understand that females can actively contribute to the economy too. The families have reduced their expectations from the women to always be saree-clad or submissive and quiet within the four walls of the house to outgoing and assertive. Women also, on the other hand, kept their familial values intact and deal in business with utmost care and self-respect.

Only one or two cases were observed where the women were running the business along with their husband. General observation was that women had their own business while the husbands had a different business or were either working. Despite being in different fields, women as well as the men, managed to devote an equal time out for themselves to enjoy their personal and family lives. Nonetheless, primary responsibility of the children is always shouldered by women. But since this was not a matter of concern, it was not considered as a hindrance in their business.

Women in business were not *solopreneurs* (a person who runs the enterprise all by themselves) but catered to a team of people working with them. Herein, the leadership abilities of respondents were put to a test. It was found that women having enterprises with greater financial turnover tend to display better leadership skills than the ones who were financially unstable. Hence, although leadership is a skill, this skill was only notably visible among the higher income groups irrespective of their age. However, to reach to a financial stability, the respondents had also given time to their enterprise as well as their team. Hence women with more work experience and greater revenue were better leaders than others. Here, the hypothesis that leadership evolves with experience was proved true.

Tabulation of dependent and independent variables also showed that Achievement Motivation, Leadership and Work life Balance, the three dependent variable of the study, are mutually correlated with each other whereas, they are not always significant with any of the Independent Variables. This means that women who have high need for achievement (n-Ach) and better leadership skills also exhibit better work life balance. This stands true for all the dependent variables.

Work life Balance is highly significant with Age of the respondents, Experience of running own enterprise and Average Income (per month). This finding emphasizes on the fact that with increased age, respondents try to find their own ways to achieve work life balance. Hence the hypothesis, Work life Balance is absent in Women Entrepreneurs (H_3) cannot be out rightly rejected as during their initial day of start-up, respondents do find it difficult to achieve work life balance, hence it is absent, however, gradually, respondents learn to strike a balance between their work life and their family life.

Work life Balance is also greatly significant with Experience of running own enterprise. With increased experience of running own enterprises, respondents find out ways how they can balance their work life and their personal life.

Work life balance is also highly significant with Average Monthly Income. This shows that as respondents start earning more, they are able to exhibit better leadership in terms of delegating and managing their work in such a manner that they can achieve better work life balance.

It can hence be strongly concluded that enterprising women are more sensitive about how their actions might affect the society around and hence they fully own the results of the decisions made by them in due course of their business. Being a woman, sensitivity, empathy and considerate nature comes naturally. Hence, enterprising women are more considerate towards their co-workers and believe in fostering a just organizational culture. However, they would not mind jeopardizing relations if any co-worker is found doing a wrong task, despite investing into training.

Hence, women entrepreneurs are as competent and assertive-aggressive as men entrepreneurs and despite the focus of this study being females only, the researcher has come to a conclusion that irrespective of the gender, entrepreneurship is a skill, which, if fostered and nurtured well and channelized in the right direction can work wonders for the economy of one family, the city, the state and the country at large.

CHAPTER VI
SUGGESTIONS
&
RECOMMENDATIONS

SUGGESTIONS & RECOMMENDATIONS

One of the objectives of the study was also to suggest a framework to promote women entrepreneurship in Gujarat. A framework for promoting entrepreneurship among women in Gujarat will be duly discussed and suggested in the next chapter.

This chapter presents the suggestions of the researcher and the recommendations along with areas of Social Work Intervention based on the findings of the analysis elaborated in the previous chapters.

This chapter is further divided in to two parts:

1. Suggestions & Recommendations
2. Areas of Social Work Interventions

1. Suggestions and Recommendations

One of the objectives for this study was to also provide a framework to develop entrepreneurship among women in Gujarat. From the study undertaken, the researcher has got many new insights about female entrepreneurship which were not known earlier. However, there are certain loopholes, to cover up which, the researcher provides suggestions as part of the study.

It is said that an entrepreneur is someone who possesses a risk taking ability and the courage to face problems and suggest alternate and novel suggestions for the same. In case of women entrepreneurs, the expectation is different because if their risk fails, they have no other choice but to shun their dreams of enterprising. It is with this perspective that the suggestions have been provided.

a. Promoting positive attitude towards entrepreneurship through role models

One of the biggest reasons women remain laid back in the field of entrepreneurship is the negative conditioning towards failure and a deep understanding that risk taking almost always leads to failure. This negativity can only be countered with a positive infusion of successful women entrepreneurs taking the lead in coaching the nascent women entrepreneurs.

- Enterprising women are different from small scale vendors and hence their psychosocial outlook is very different. Their education and skills can be effectively used for creating proper forums for budding entrepreneurs wherein the stable entrepreneurs can explain the challenges and mentor the budding entrepreneurs to nurture better.
- The dedicated forums for women entrepreneurs can offer periodical trainings which can cater and invite more and more participation from enterprising women in order to hone their entrepreneurial skills.
- Voluntary role-model engagement programmes can be encouraged where successful women entrepreneurs come together at public events to encourage entrepreneurship among women.
- These forums can be utilised to incubate budding entrepreneurs wherein the skill development can be rigorously emphasised.
- These forums can also act as a platform for women entrepreneurs to effectively conceptualise their entrepreneurial ideas and convert it in to successful businesses.

b. Approaching the proper authorities

- There is an increase in the emphasis of promoting women entrepreneurship under the aegis of Skill India Program of the Government of India and the state government as well. However, there are many entry points for women and many times this creates a confusion among those who are actually seeking help.

- It is suggested to have a single-window system for the registration to loans to scholarships and benefitting schemes, and other aspects related to entrepreneurship. Here, it becomes important to understand that women entrepreneurs may not always have a self-help group (SHGs) or support groups, they may be individuals willing to start up a firm on their own. For such women, a separate system must be devised without them having to run from pillar to post to ensure their paperwork is in place.
- Use of the internet and online systems can prove to be a great boon to the government functionaries and the beneficiaries alike in order to streamline processes for every aspect of entrepreneurship.
- Enterprises registered by women in the name of women must be specially recognised and some rewards can be provided to those enterprises. This strategy may encourage other women also to come up with novel ideas and can lead to increased entrepreneurship.

c. Facilitate access to finance through financial literacy and a range of financing instruments

- There are various financial schemes available for women entrepreneurs in India but the beneficiaries are unaware about it because e of lack of awareness and proper advertisement.
- It is observed that women entrepreneurs lack financial literacy and that financial institutions are also not very open about the explanation regarding the complicated policies.
- Modern methods like "fintech" as an industry sector is changing the nature of business funding, and small businesses are increasingly using a wider range of sources of capital.
- Use of Crowdfunding is also becoming more popular and is applied in many forms, including rewards or gifts, debt, equity, or a combination of debt and equity. Crowdfunding assures of reaching out to a large network of potential financiers over the

internet without any face-to-face contact. By resorting to these modern methods, female entrepreneurs may be able to overcome the traditional discrimination in accessing financial backing to start a new ventures. (Halabisky, Potter, Garcia, & Levie, 2016)

d. Easing journey from the mind to the market

- Entrepreneurship is all about the right ideas at the right time combined with the right resources and we have a successful enterprise. However, when it comes to enterprising women, a lot of ideation remains limited only within the enclosures of their minds. The scepticism of failing does not let women take many risks. Moreover, the linkages between the production and marketing of the same are found to be very weak and in their nascent stage.
- Inability to promote their own products or services leaves women dejected and withdrawn from entrepreneurship. This can be explained as every woman knows how to cook, but not every woman is a restaurateur.
- In Gujarat, a government arm of the Mission Mangalam - Gujarat Livelihoods Promotion Company (GLPC) Limited has initiated exhibitions where women entrepreneurs can rent out stalls to market their products. However, these kind of initiatives need to be increased and promoted at a higher level so as to cater to a larger population.

e. Promoting Work life Balance and Family Counselling

- The social roles provoke women to ensure the house is kept in proper order while the man is at work. However, this cannot be convincingly said in case if the women begin working and leave the house for the significant other members to take care of. Hence, sensitization within families needs to be created so that they can promote women in their entrepreneurial activities.
- Young female entrepreneurs also have the responsibilities of marriage and family due to which their entrepreneurial role takes

a back seat. A supportive and understanding family can help women overcome their inhibition even during these life events.

- Sensitization to explain that women can also portray the role of an entrepreneur must be created right from the beginning.
- Schools and colleges must adopt Entrepreneurship as part of their curriculum so that idea generation can start at a very young age irrespective of the gender and that deeper family understanding is created to ensure proper support.

2. Areas of Social Work Interventions

The accepted Global definition of Social Work Profession defines that "Social work is a practice-based profession and an academic discipline that promotes social change and development, social cohesion, and the empowerment and liberation of people. Principles of social justice, human rights, collective responsibility and respect for diversities are central to social work. Underpinned by theories of social work, social sciences, humanities and indigenous knowledges, social work engages people and structures to address life challenges and enhance wellbeing." (International Federation of Social Workers, 2014)

Hence, the professional social work respects diversity and enhances well being by use of various methods by which an individual, group or community at large can be empowered to help itself even in absence of any individual or situation. The gamut of social work has always focused its attention on the weaker and vulnerable sections of the society and tried to mainstream them by countering the difficulties and enhancing welfare.

Through the current study also, the researcher has tried to bring about interventions which can augment entrepreneurship among women and help them be self-reliant.

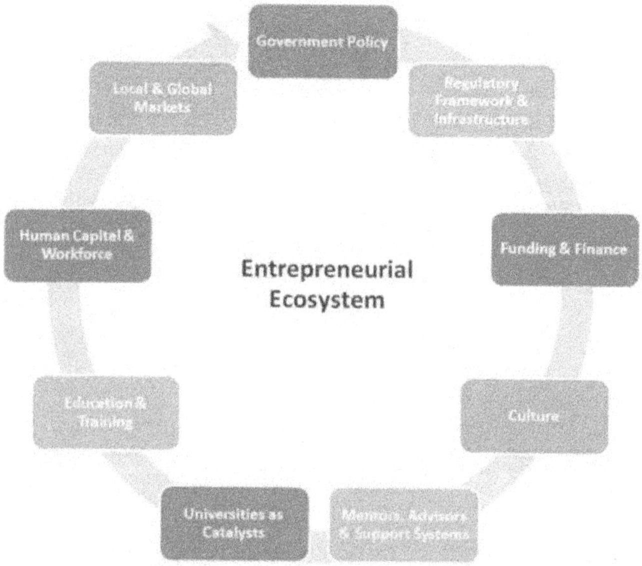

Figure 6.2.1

Showing Entrepreneurial Ecosystem

Figure 6.2.1 briefly explains the entrepreneurial ecosystem in which there are a lot of factors which play their role on helping the entrepreneur rise above their inhibitions and become a successful business-owner. Each of these elements of the entrepreneurial ecosystem can be linked with Social Work Methods to prepare a proper framework for intervention.

Social Work has two methods:

1. Primary Methods
    - 1.1. Social Casework
    - 1.2. Social Groupwork
    - 1.3. Community Organization
2. Secondary Methods
    - 2.1 Social Welfare Administration
    - 2.2 Social Work Research
    - 2.3 Social Action

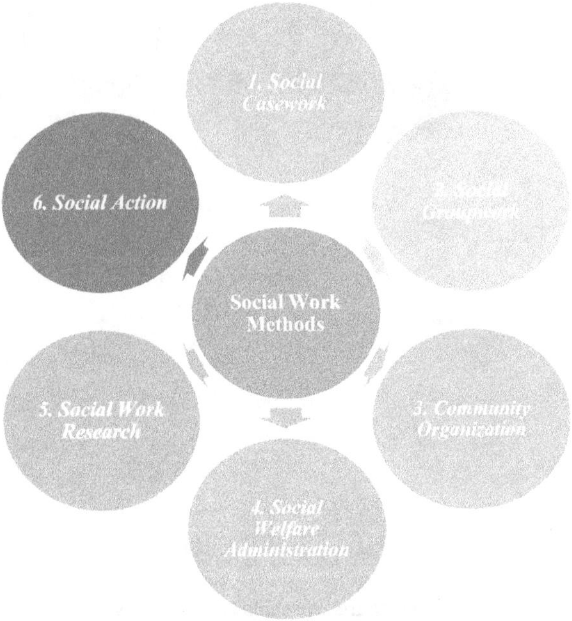

Each method is unique in its approach to ease out human lives by providing scope for improvement at every aspect. Each aspect is described in detail below with areas for Social Work intervention at various levels to augment and accelerate the growth of women entrepreneurship in the state and the country at large.

1. Social Casework

Mary Richmond (1917) defined Social Casework as "the art of bringing better adjustment in social relationships of men, women or children." The purview of casework is dealing with individual problems and help them overcome it.

By applying the theories of casework to the current study, women can be empowered with greater confidence within themselves. In the Indian context, family dependency and personal obligations on women create hindrances to them in succeeding in their career. Motivational factors: Factors like intrinsic motivation, government policies, financial assistance and family support are still in weak shape in India for women entrepreneurs. These individual issues can be taken up and resolved in order to facilitate women. Providing proper hand-holding until these women are fully confident in the way they carry themselves in the market.

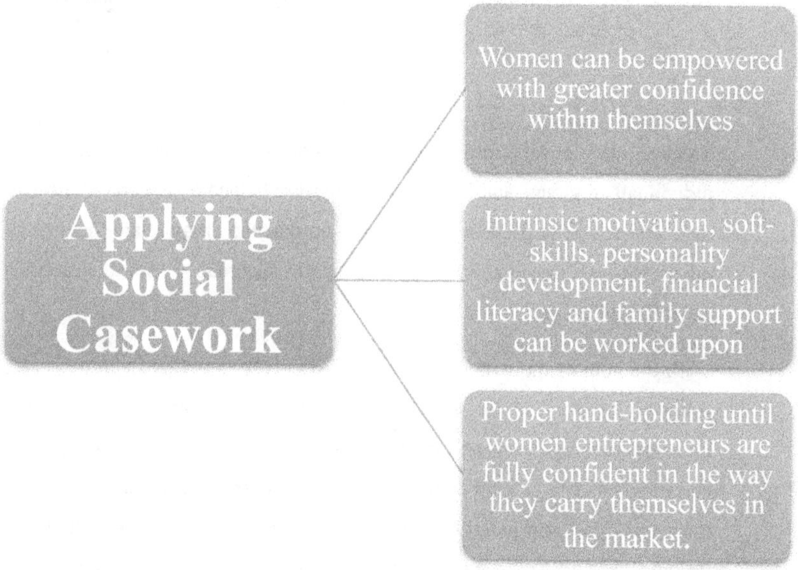

2. Social Groupwork

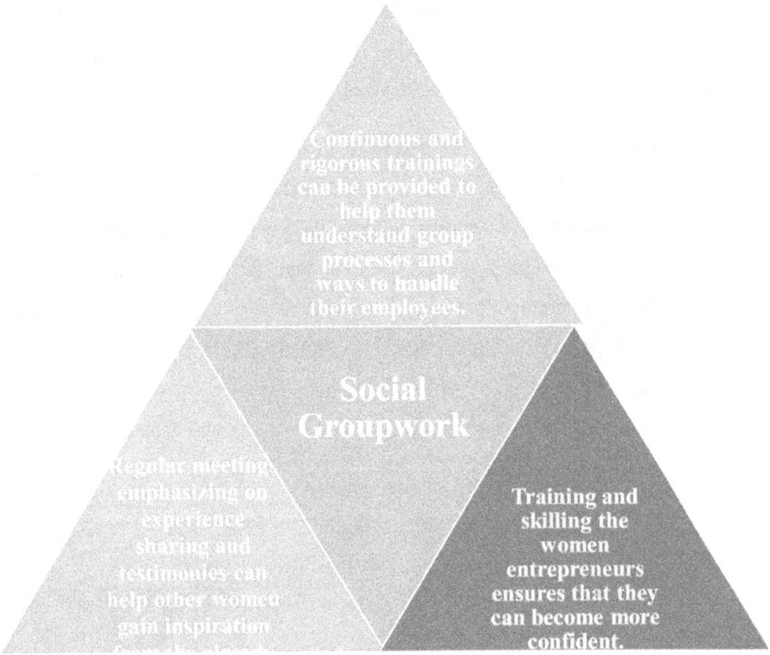

Trecker (1955) defined Social Groupwork as "a method through which individuals in groups and in agency settings are helped by a worker who guides their interaction in programme activities so that they may relate themselves to others and experience growth opportunities in accordance with their needs and capacities to the end.

With the help of the method of Social Groupwork, a focused approach can be followed to work in the direction of empowering women. Continuous and rigorous trainings can be provided to a small group of budding women entrepreneurs and helping them understand group processes and ways to handle their employees and practice leadership. With constant and consistent efforts in training and skilling the women entrepreneurs, they can be made to become more confident as a group.

Moreover, regular meetings with groups of women emphasizing on experience sharing and testimonies can help other women gain inspiration from the already successful women.

3. Community Organization

Ross (1967) defines community organisation as a "process by which a community identifies its needs or objectives, gives priority to them, develops confidence and will to work at them, finds resources (internal and external) to deal with them, and in doing so, extends and develops cooperative and collaborative attitudes and practices in the community".

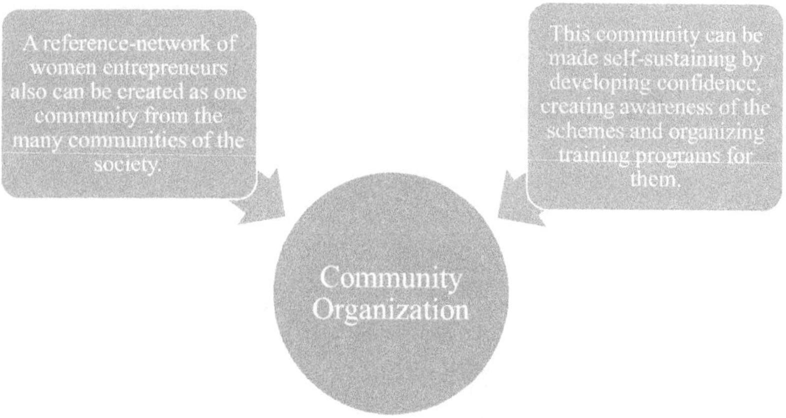

A reference-network of women entrepreneurs also can be considered as one community from the many communities of the society. This community can be made self-sustaining by developing confidence in them, creating awareness of the various schemes which can provide financial assistance and organizing training programs for them.

Moreover, the various methods of community organization like Education promotion to promote entrepreneurial education, Legislative promotion to promote various schemes offered by the government organizations, non-legislative promotion to promote the schemes offered by voluntary organizations, fund raising for enterprises, conferences of reference networks and evaluation of the status of enterprises can also be done for the smoot functioning of enterprises.

4. Social Welfare Administration

Harleigh Tracker (1971) interprets social welfare administration as a "process of working with people in ways that release and relate their energies so that they use all available resources to accomplish the purpose of providing needed community services and programmes." (Devi & Ravi, 1998)

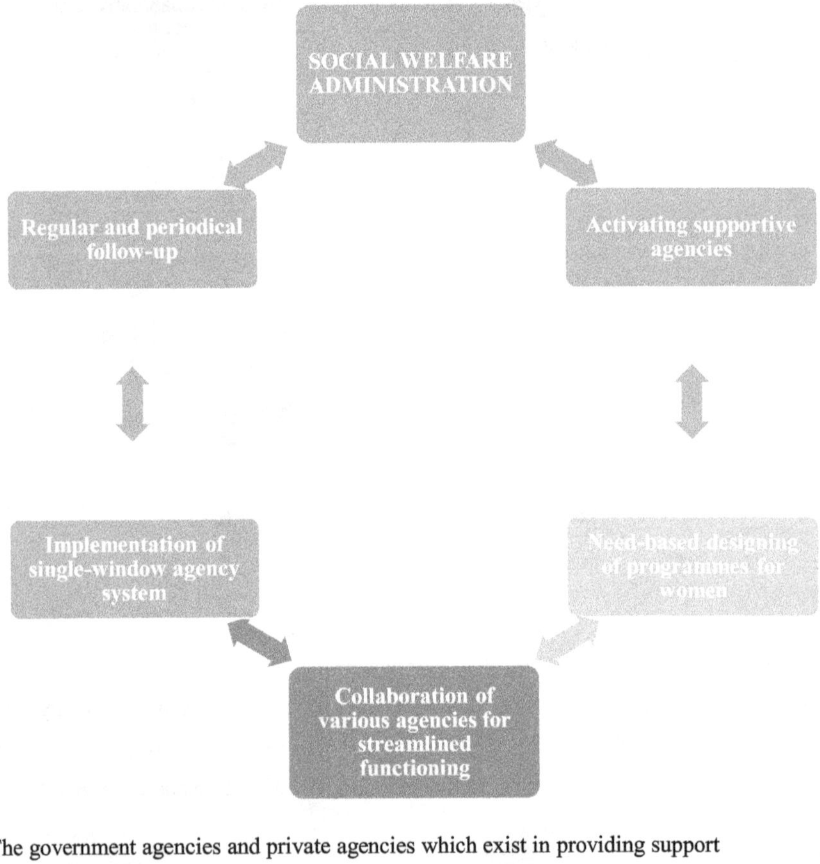

The government agencies and private agencies which exist in providing support to the enterprising women can be made more active in their functions so that a regular follow-up on trainings is taken and need-based programmes are designed for the benefit of the women entrepreneurs. The District Industries Centers (DICs), the Centres for Entrepreneurship Development, Chambers of Commerce and Industries, national and cooperative banks and all other significant agencies should work in collaboration in order to properly assist women in their trade and business guidance. A single-window agency system must be implemented wherein women can get their issued resolved under one single roof.

5. Social Work Research

Social work research may be defined as systematic investigation into the problems in the field of social work. (Das D. K., Practice of Social Research : A Social Work Perspective, 2000)

Social Work Research		
Impact of government schemes Impact of financial literacy Market linkages and women	Rural – Urban Entrepreneurship Micro-financing Sector-wise presence of Women Entrepreneurs	Gender analysis of entrepreneurship

There is a great scope of researching in the field of women entrepreneurship from a social work perspective. The researcher has made a humble attempt in identifying and studying certain psychosocial issues of entrepreneurship among women but many areas such as impact of government schemes, impact of financial literacy, market linkages and women, leadership, gender analysis of entrepreneurship can be undertaken to gather insights from the field and can be researched upon.

Research in any topic will yield to generation of further scope of study. Similarly, the current study though has limitations, yet many researches can be formulated on the lines of the same topic as that under current study.

6. Social Action

Singh (1986) explains that social action is a process in which conscious, systematic and organised efforts are made by some elites and/or people themselves to bring about change in the system which is instrumental in solving problems and improving conditions which limit the social functioning of weaker and vulnerable sections. (Mishra P. D., 1992)

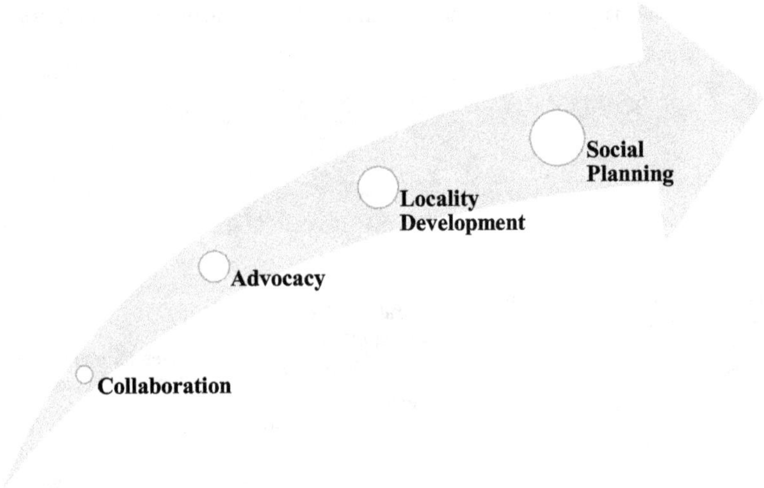

It is only by collaborating the various efforts made by the various agencies that something positive in the direction of helping women entrepreneurs grow and develop well can be achieved. Proper use of advocacy of the rights of women and women entrepreneurs can help foster the area of entrepreneurship among women. Models of Locality Development and Social Planning can be effective in practicing Social Action within the entrepreneur community.

Thus by working on these aspects, a proper intervention strategy for empowering women through entrepreneurship can be made.

From the above, it can be summarized diagrammatically that development of women entrepreneurship basically requires certain common and collective actions to be undertaken. If designed and implemented well, these actions can pave a new road towards achieving and equal society for women entrepreneurs. These actions are:

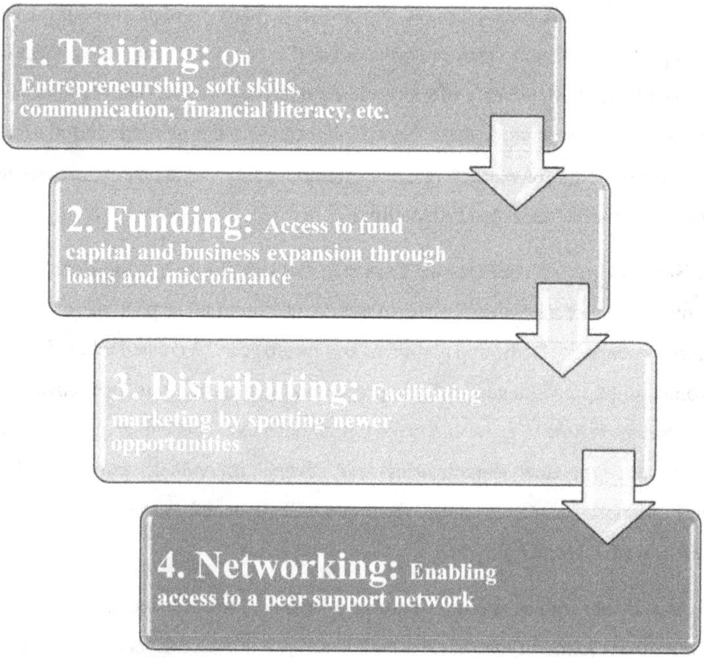

1. Training: Effective trainings can be planned and arranged for focused groups of women entrepreneurs wherein they can be trained on various entrepreneurial aspects like Proposal Formation, Soft Skills, Communication, Financial Literacy, Banking Procedures, Government Schemes available for support, collaborating and networking as well as overall entrepreneurship development. A proper handholding mechanism must be designed which helps women overcome their diffidence.

2. Funding: Proper linkages must be created between financing agencies and the beneficiaries. Technology and crowdfunding can also be used to its fullest potential to encourage women entrepreneurs. Understanding the need of the business and learning to respond to it by expanding the capital investment with the help of loans and other means of financial support must be encouraged among women. Awareness of these facilities must be spread through proper media so that the number of beneficiaries may rise.

3. Distributing: Effective market linkages must be created in order to facilitate and encourage greater participation from women entrepreneurs. Due to lack of avenues to market their products and services, many women give up on their entrepreneurial dream. This can be avoided if a facilitating market connect can be established with the help of appropriate authorities. Internet has revolutionised the way businesses work these days. Understanding the using digital platforms and social media for marketing can be introduced so that women entrepreneurs can easily market their products online.

4. Networking: Businesses are known to flourish if one has the right kind of contacts. Hence networking becomes a very important aspect of developing entrepreneurship. Dedicated platforms and forums can be created where budding women entrepreneurs can easily approach and get proper guidance, advice and mentoring on how to grow and expand their business. This can enable a learning ecosystem for the beneficiaries and every individual, experienced or inexperienced in managing their business will be benefitted thereby creating a win-win situation for all.

As the society is undergoing a major paradigm shift towards technology, there is requirement to modify the society beliefs and provide women entrepreneurs with a strong ecosystem for entrepreneurship and democracy.

On a concluding note, the researcher would like to quote a very famous poem from one of the very well known and renowned Afro-American Women's Rights Activist, Maya Angelou (Angelou, 2011) on women. The focus and context of her work was basically expressing the extraordinary characteristics of women, however, the researcher has taken liberty to change some words in order to make it more meaningful and apt in the context of the current study.

Home makers wonder where my secret lies

Working women question my business size

But when I start to tell them,

They think I'm telling lies.

I say, it's in the reach of my arms,

the span of my waist

the stride of my step

the curl of my hair, in my fingers, interlaced

I'm a Woman. Phenomenally.

Phenomenal Woman. That's Me.

Inspired by
Maya Angelou

Women Entrepreneurs: A Study of Achievement Motivation, Leadership and Work-life Balance of Women Entrepreneurs of Gujarat

Introduction

According to P. F. Drucker an entrepreneur is one who always searches for change, responds to it and exploits it as an opportunity."

According to Cantillion "entrepreneur is the agent who buys means of production at certain prices, in order to sell at prices that are certain at the moment at which he/she commits himself/ herself to his/ her cost".

There exist a number of schools of thought, which view the notion of entrepreneurship from fundamentally different perspectives. The term has been used to define a wide range of activities such as creation, funding, adapting, and managing and venture.

Recently a new trend has emerged in India wherein women are venturing as entrepreneurs and are contributing to the economic development of the country. Women are increasingly seeking entrepreneurship as an avenue for economic growth. The government is also playing a significant role in mobilising women to become entrepreneurs through its different programs. This is helping is the number of women entrepreneurs increase much more.

Women Entrepreneurs

The concept of women entrepreneurship is becoming a global phenomenon. Women access to knowledge, skills, resources, opportunities and power but still remain rather low especially in the rural areas. Although women remain involved in family trade but they are neither in organized form nor do women have its ownership making their role as a mere helper. Women of today are clearer in their life-goals. They know how to undergo all the hardships as well as how to cherish the success which comes after. In our culture, women are socially conditioned to be submissive and not speak much. But a woman with entrepreneurial aspirations must make herself heard. While making tall claims on achieving women empowerment, we must also provide opportunities and the right kind of platforms to women to achieve.

Women have been taking increasing interest in recent years in income generating activities, self employment and entrepreneurship. This is seen in respect of all kinds of women both in urban and rural areas. Women are taking up both traditional activities and also non-traditional activities. It is clear that more and more women are coming forward to set up enterprises. In the process of entrepreneurship, women have to face various problems associated with entrepreneurship and these problems get doubled because of her dual role as a wage earner and a homemaker. It becomes all the more important to study what keeps these women well motivated in order to achieve their set goals and targets. It is also important to study what leadership patterns are followed by these women in order to handle their human resources and other resources as well as how they balance their work-life while setting up their enterprises.

Social Work Significance

The profession of social work has always emphasized on empowering women and it is considered to be one of the most important thrust areas of the profession. Encouraging livelihoods and promoting skill-development has always been of prime importance to the field of social work. Every trained social worker aims to uplift the status of their stakeholders in every possible way. Promoting gender equality and encouraging gender-neutral workplaces and work environment is the primary principle of social work practice. With the dawn of the new century, women have sought to look more beyond just a stable job. The concept of women entrepreneurship is becoming a global phenomenon. Women access to knowledge, skills, resources, opportunities and power but still remain rather low especially in the rural areas. Although women remain involved in family trade but they are neither in organized form nor do women have its ownership making their role as a mere helper. Women of today are clearer in their life-goals. They know how to undergo all the hardships as well as how to cherish the success which comes after. In our culture, women are socially conditioned to be submissive and not speak much. But a woman with entrepreneurial aspirations must make herself heard. While making tall claims on achieving women empowerment, we must also provide opportunities and the right kind of platforms to women to achieve.

An entrepreneur plays a very dynamic role for his enterprise. A woman, considered vulnerable, has to combat a lot of hurdles and obstacles which come in her way achieving her goals. The researcher will also consider the environmental challenges, strengths and weaknesses of women entrepreneurs and that how under even stressful times, a woman exhibits strength in holding her career and personal life together.

Duality of the roles of Women

Women entrepreneurs have to play a dual role in their lives. They are supposed to undertake household chores as well as manage the enterprise. If a woman seems to be too ambitious towards her career, she is labelled as insensitive towards her homely responsibilities. Women forget that they too have a capacity in trying to fulfil their dual roles. This feeling of being a super-woman, gives way to conflicts. Career women often have to face work-home role conflict and so the concept of work-life balance becomes an important parameter to be studied while researching on women entrepreneurs.

Working women and especially enterprising women need to be authoritative and showing some leadership skills. It also becomes necessary to study the leadership aspect of enterprising women as that factor helps them retain manpower and expand their enterprise. The researcher will also study the leadership styles adopted by enterprising women and how it affects their entrepreneurial performance.

Women entrepreneurs represent a group of women who have broken away from the beaten track and are exploring newer avenues of economic participation. Women with high drive, creativity, innovation and those who are keen on achieving on their own have taken up the challenging role of entrepreneurship. Women no longer confine themselves to conventional fields like embroidery, knitting and tailoring but are now venturing into new fields like electronics, pharmaceuticals engineering and services. These women are mostly urban, appropriately educated, sometimes having specialised skills and equipping themselves with some training and financial support. In spite of women taking entrepreneurship in many challenging fields the current entrepreneurial activity in India is not very high. There exist a definite gap between the current status of women entrepreneurial activity and their potential. The social origins of women entrepreneurs play an important role in defining the kind of work they undertake. Women also find micro enterprises attractive because of low barriers to entry and the flexible nature of work which makes it easy to combine gainful employment with domestic responsibilities. (Carr, 1990)

Women Entrepreneurs in Gujarat

Business runs in the blood of every Gujarati. It has spread its acumen over USA, Canada and many other Western horizons. The Centre for Entrepreneurship Development (CED) was established in 1970 in Gujarat to enhance trader's expertise and help the businesses flourish. In order to encourage more women entrepreneurs from the state, CED started various programmes to train women in this domain.

The Government of Gujarat has taken many steps to encourage the women of the state to take up their entrepreneurial careers. While he was the Chief Minister of the state, Shri Narendra Modi kept on emphasizing in his speeches that entrepreneurship is inbuilt in the people of Gujarat. He further kept addressing that women play an important role for the economic development. According to him, even in the dairy sphere, the contribution of men is minimal. He suggested that women should be made an integral part in the decision making process in almost all the spheres.

High need for achievement, high need for autonomy, high need for creativity, and prevalence of initiative, industriousness, and foresight through self-reliance and possession of managerial skills are all effective pointers to an entrepreneur. (Sarada, 2004). In this globalised and digitalised era, many women have found their place in the market as entrepreneurs. With the changes in the way a business runs, it becomes important to study how motivated women are to pursue their dream of enterprising. There is a great scope to study the work-life balance and leadership skills of the women of the 21st century as this era is an era of continual transformation and change and entrepreneurs have to work up to their mark and even beyond to survive in an otherwise competitive market. The Government also has rolled out the "Start-up India" initiative to encourage more and more entrepreneurs in the various business sectors.

Social Work is an inclusive profession. Its core values speak of inclusive development and acceptance of people. Women have always borne the brunt of social changes and therefore labelled as vulnerable. However, in this fast-paced world, the challenges are of a different kind. It has been a long pending economic independence that women have achieved on their own merit. Former leaders of the country like Mahatma Gandhi and Dr. B. R. Ambedkar had always encouraged participation of women in the freedom struggle and economic development of the country. Despite this ideology, it took years together for women to achieve this freedom. In recent years, women have started being more and more assertive about their economic independence and many such women are turning to entrepreneurship. They are now their own bosses. Women may have reached to the pinnacle of success but not all success comes easy. Due to the dual social roles that women play, success would not have been possible without sacrificing their family life or other social expectations. It is with this intention that this study needs to take place so that the expectations, ideas and aspirations of the enterprising women of the millennium can be understood and appropriate assistance can be provided to foster their skill, capacity and talent.

Objectives of the study

1. To study the socio-economic background of women entrepreneurs in Gujarat
2. To analyse motivational factors behind women entrepreneurship
3. To analyse the leadership quality among women entrepreneurs in Gujarat.
4. To analyse the degree of work-life balance that women entrepreneurs face and its effect on their entrepreneurial performance.
5. To suggest a framework to promote women entrepreneurship in Gujarat.

Operational Definition:

Study title - Women Entrepreneurs - A Study of Achievement Motivation, Leadership and Work-life Balance of Women Entrepreneurs of Gujarat.

Women Entrepreneurs - Women entrepreneurs represent a group of women who have broken away from the beaten track and are exploring newer avenues of economic participation. (Rani, 1996)

Achievement Motivation - The concept of achievement motivation has been used to describe individual efforts towards personal goals in social surroundings (Cassidy & Lynn, 1989). It is also found to be a typical characteristic among individuals who are successful in their work (McClelland, The Achieving Society, 1961) (McClelland, Money as motivator: some research insights., 1967). A strong achievement motivation is probably the one characteristic most clearly describing the attitudes and the behaviour of an entrepreneur (Virtanen, 1997).

Leadership - Leadership is a process whereby an individual influences a group of individuals to achieve a common goal. Leadership style is characterized by the consistent patterns of exhibited behaviors, attitudes, and values. Leadership has been defined as traits, behaviors, influence, persuasion, interaction patterns, role relationships, and as administrative position; and most definitions assume that leadership involves a process whereby intentional influence is exerted by one to others in an attempt to guide relationships (Moerer-Urdahl, 2005).

Gender is integral to contemporary notions of effective styles that have morphed from a traditional masculine, autocratic style to a more feminine and androgynous style of democratic and transformational leadership (Knopik & Moerer, 2008).

Work-life Balance - Work-Life Balance does not mean an equal balance. It means the capacity to schedule the hours of professional and personal life so as to lead a healthy and peaceful life. It is not a new concept. It emphasizes the values, attitudes and beliefs of women regarding their age to work in organizing and balancing their work and personal life. (Deshmukh & Deshmukh, 2016)

Women turn to entrepreneurship for various reasons especially for satisfying their personal needs of self-expression an autonomy. This also makes their lives increasingly complex with pressure, intense working hours and lesser involvements in the family; hence they juggle to satisfy both the roles. (Rani, 1996) They may receive additional help at home in the form of servants, parents, in-laws and husband but still the entire responsibility of the home rests with the woman. But the problems remain, as the entrepreneur is not a super woman and there follows a conflict.

Relevance of this study to Social Work Profession

Social Work is an inclusive profession. Its core values hold inclusive development and acceptance of people. Women have always borne the brunt of social changes and therefore labelled as vulnerable. However, in this fast-paced world, the challenges are of a different kind. Women may have reached to the pinnacle of success but not without sacrificing their family life or other core values. It is with this intention that this study needs to take place so that the expectations, ideas and aspirations of the enterprising women of the millennium can be understood and appropriate assistance can be provided to foster their skill, capacity and talent.

Research Methodology:

Who is a woman entrepreneur? This question has been defined in different ways by different people at different points of time. The Government of India considered the majority participation of women in enterprises as a necessary qualification for calling such enterprises as women-enterprises. Recently, the term majority participation has been modified to majority equity participation of women in organizations. This again, could be debated as many times, women may invest but not as actively participate in the running of the enterprise. It has been thus thought that both equity participation as well as participation in the running of enterprises to a greater extent by women entrepreneurs should be a necessary criteria for calling enterprises as women-run enterprises.

Therefore, for this study, the women entrepreneurs considered were based on the following **inclusion criteria**:

1. The woman entrepreneur should have contributed to the equity capital of the concern.

2. The woman entrepreneur should be actively participating in the running of the organization and in all functional areas of management.

3. The establishment of the enterprise be minimum three years.

4. The establishment must help generate income for all those employees associated with it.

Discussions with appropriate authorities at the government departments like District Industries Centres (DICs) and Centres for Entrepreneurship Development (CEDs) as well as the Chambers of Commerce and Industries and Entrepreneurship Development Institute (EDI) were done. With the preliminary information provided, an attempt to formulate a list of women-run enterprises was created. It was felt that as no census data is available on the number of women entrepreneurs in the state, **a snowball sampling method** was followed.

This study follows **an exploratory cum descriptive research design,** as it will help the researcher to understand the issue more thoroughly, before attempting to quantify mass responses into statistically inferable data. Used properly, exploratory research will provide rich quality information that will help identify the main issues that should be addressed in our survey and significantly reduce the level of prejudice. Within the exploratory design, the researcher will incorporate the use of Case studies. One case study from each zone was presented. Reliability on Secondary Sources of Data was also adopted.

Tools of Data Collection:

A properly structured tool for collecting demographic data was constructed using more of close-ended questions as the study aims at obtaining the opinions of its respondents. This tool was developed by referring the various standardised tools available on the relevant subject. A pilot sample was studied and the sample data was treated on Cronbach's Alpha Test to assure of the reliability and validity of the tool.

Universe:

For the purpose of this study, the researcher studied all the women-owned, women-run enterprises within the State of Gujarat. However, four major zones i.e. Ahmedabad, Rajkot, Surat and Vadodara were covered.

Sampling:

Snowball Sampling method was used. In this method of sampling, the existing subjects under study refer the future subjects from among their acquaintances. The research scholar tried to obtain the names of Women Entrepreneurs from the local Chamber of Commerce and Industries from each of the zones under study and proper representation from each zone was selected for the study. All women-owned, women-run enterprises established since the last three years in the cities of Ahmedabad, Rajkot, Surat and Vadodara were studied.

Variables:

Dependent Variables: Work-Life Balance, Motivation and Leadership

Independent Variables: Age, Education, Experience and Income

Hypotheses: 1. Higher the income, higher is the motivation to work.
2. With increased experience in the field, entrepreneurs develop their own leadership.

3. Work-life balance is absent in women entrepreneurs.

Statistical Treatment:

The filled questionnaires were numbered serially and checked for consistency of the data, and any in-consistent data. The questions and responses were codified to represent the data in tabular / graphical form wherever appropriate. The collected data was correlated with the various dependent and independent variables. Hypotheses were verified using appropriate statistical tools and with the help of Excel / SPSS (Statistical Package for Social Science for Windows).

Limitations:

According to the researcher, no research can be a perfect research. There is always a scope of error. In this research, the researcher feels that there may be certain limitations that might affect the study. These limitations are,

- Lack of willingness of the respondent to participate in the research.
- Non-availability of Respondents.
- Lack of understanding about the topic in concern by the respondents.
- Authenticity of the secondary sources of data.

Data Analyses

The structured tool was first treated under the Cronbach's Alpha Test to find out its reliability and validity. On the exposed sample of 30 tools, the following score was found:

Case Processing Summary			
		N	%
Cases	Valid	30	100.0

	Excluded[a]	0	.0
	Total	30	100.0

a. List-wise deletion based on all variables in the procedure.

Reliability Statistics

Cronbach's Alpha	Cronbach's Alpha Based on Standardized Items	No of Items
.849	.850	55

As per the obtained score, it can be said that the designed tool was highly reliable for its use and it would measure what it intends to measure.

From the analysed data on 15 tools, the following has been inferred:

Socio-economic Profile

- 59 (37.3%) respondents fall between the age groups of 31-40 years.
- 71 (44.9%) respondents have pursued education up to Graduation level whereas 52 (32.9%) of the respondents have pursued education up to Post Graduate level.
- Majority of the respondents i.e. 77.2% are married.
- 54 (34.2%) respondents have an experience of more than 10 years in running an enterprise.
- 60 respondents, i.e., 38% have an experience of 2-5 years in running their current enterprise whereas 38 (24.1%) respondents have an experience of running their own enterprise for the last 6-10 years.
- 41 respondents, i.e., 25.9% have an average monthly income of Rs. 21,001 - 50,000. A sizeable number of respondents, 32 (20.3%) have answered that their monthly income is not regular.

- 43 respondents' husbands have their own business i.e. 27.2% and an equal per cent belong to professional category. Professions like Chartered Accountants, Doctors etc. are considered a part of the Professional category.
- While 39 respondents, i.e., 25.3% have chosen not to disclose their husband's average monthly income, 24.7% respondents have claimed that the average monthly income of their husbands is more than Rs. 50,000/-.
- 57 respondents, i.e., 36% have husbands whose minimum qualification is graduation.
- Majority of the respondents, i.e., 81.6% have 3-5 members in their families.
- 51 respondents, i.e., 32.3% have only one child.

On the scale of **Achievement Motivation**, the following inferences were made -

Table 2.1.1 Cumulative response on Achievement Motivation

N=158

Achievement Motivation		
	Frequency	**Percent**
Mostly disagree	15	9.5%
Agree to some extent	90	57%
Mostly agree	53	33.5%
Total	**158**	**100%**

The above graph shows the cumulative responses of the respondents on Achievement Motivation. It shows that out if the 158 respondents, majority of the respondents, i.e., 57% agree to some extent on having Achievement Motivation, 33.5% mostly agree that they have high Achievement Motivation and 9.5% of respondents disagree on having Achievement Motivation.

Graph 2.1.1 Showing Achievement Motivation of Respondents

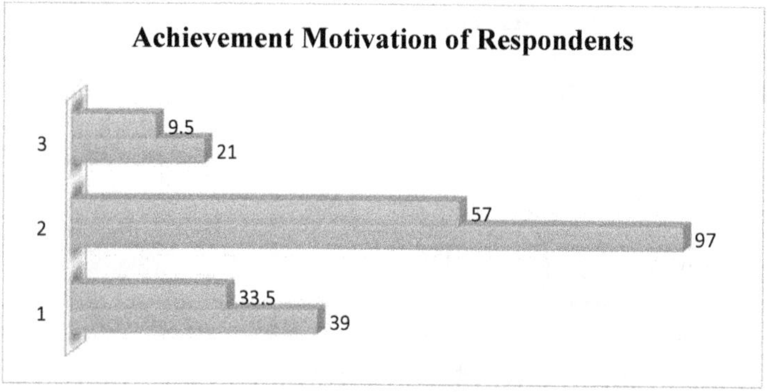

Thus, it can be concluded that there is visible level of achievement motivation present among the respondents. It is this motivation that aspires women to work harder and achieve greater heights in their career. There is a great sense of achievement in women, and on knowing that people like them have achieved much in life, they are also motivated to work harder.

This also hints at the aspiring and hardworking nature of the women respondents.

Respondents own their actions and believe that they themselves like to deal with difficult situations so that the blame or praise comes to them alone.

On the scale of **Leadership**, the following trends were observed -

Table 3.1.1 Showing Cumulative responses on Leadership

N=158

	Frequency	**Percent**
Sometimes	39	24.7
Always	119	75.3
Total	**158**	**100.0**

Since none of the responses hint towards negation in Leadership, the 'Never' option has been deleted for easy understanding, The above table can be graphically represented as:

Graph 3.1.1 Showing Cumulative responses on Leadership.

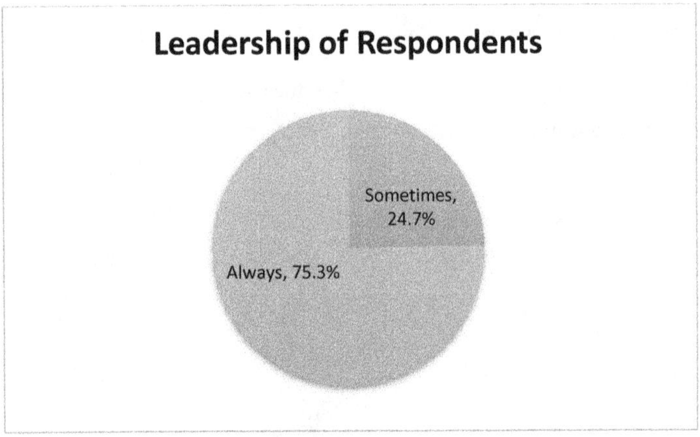

From the data, it can be said that most of the women accept that they closely monitor the tasks at hand.

Many respondents believe in understanding and respecting other peoples' boundaries and limitations thereby knowing how to delegate tasks to them. This also makes them efficient leaders as they understand and accept their team members as they are. A more androgynous cum transformational leadership style is visible among women entrepreneurs.

Respondents feel they need to be more connected with their staff and hence may overlook mistakes however, if the mistakes can lead to serious consequences, they do not hesitate in putting their personal relations apart from their professional ones.

On the scale of **Work-life Balance**, the following inferences were made -

Table 4.1.1 Showing cumulative responses of Work life Balance of Respondents

N=158

	Frequency	**Percent**
If it little true	01	0.6
If it is somewhat true	39	24.7
If it is fairly true	97	61.4
If it is definitely true	21	13.3
Total	**158**	**100.0**

The above table shows that out of the total respondents, i.e. 158, for a majority of respondents, 61.4% achieving work life balance is fairly true, for 24.7% respondents, achieving work life balance is somewhat true, for 13.3% respondents, achieving work life balance is definitely true whereas for only a small population of 0.6%, it is difficult to achieve work life balance.

This can be more descriptively represented through a graph, as below:

Graph 4.1.1 Showing cumulative responses of Work life Balance of Respondents

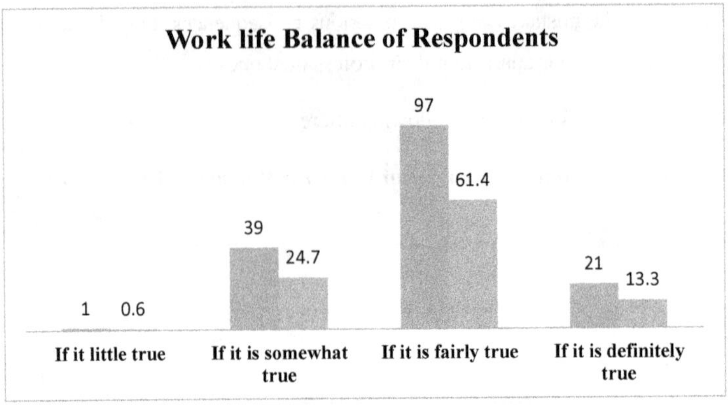

As far as Work-life Balance is concerned, the respondents felt that their personal lives were not affected by their work and they could easily balance it out.

Respondents expressed that they felt a deep sense of satisfaction by contributing to their family and the society at large by being in business and this feeling motivated them to push harder to achieve their goals.

On the basis of the data collected and statistically treated statistical, a collective result on the overall association between Achievement Motivation, Leadership and Work life Balance of Respondents was prepared.

A tabular representation of the same is as below:

Table 5.13 Showing Association between Independent and Dependent Variables.

N=158

		Achievement Motivation	**Leadership**	**Work Life Balance**	**Age of the Respondent**	**Education**	**Experience**	**Average Income (Per Month)**
Achievement Motivation	Pearson Correlation	01	.456**	.367**	.095	.055	.001	.003
	Sig. (2-tailed)		**.000**	**.000**	.235	.493	.994	.974
	N	158	158	158	158	158	158	126

Leadership	Pearson Correlation	.456**	01	.422**	.011	.077	.062	.301**
	Sig. (2-tailed)	**.000**		**.000**	.890	.334	.441	**.001**
	N	158	158	158	158	158	158	126
Work Life Balance	Pearson Correlation	.367**	.422**	01	.197*	.028	.168*	.376**
	Sig. (2-tailed)	**.000**	**.000**		**.013**	.723	**.035**	**.000**
	N	158	158	158	158	158	158	126

(The mean difference is significant at the 0.05 level.)

The above table shows association between the independent and dependent variables. It can be seen that Achievement Motivation is highly significant and associated with Leadership and Work life Balance whereas, it is not significant with any of the Independent Variables.

Leadership is highly significant and associated with the dependent variable of Work life balance and the Independent variable of Average Income (Per Month). It can be interpreted that respondents who have higher income have better leadership skills. However, there is no significance between leadership and experience. Hence, the hypothesis that with greater experience, the respondents develop better leadership skills is rejected. This gives scope to study the leadership patterns of respondents and what factors foster leadership among women entrepreneurs.

Work life Balance is highly significant with Age of the respondents, Experience of running own enterprise and Average Income (per month). This finding emphasizes on the fact that with increased age, respondents try to find their own ways to achieve work life balance. Hence the hypothesis, Work life Balance is absent in Women Entrepreneurs (H_3) cannot be out rightly rejected as during their initial day of start-up, respondents do find it difficult to achieve work life balance, hence it is absent, however, gradually, respondents learn to strike a balance between their work life and their family life.

Work life Balance is also greatly significant with Experience of running own enterprise. With increased experience of running own enterprises, respondents find out ways how they can balance their work life and their personal life.

Work life balance is also highly significant with Average Monthly Income. This shows that as respondents start earning more, they are able to exhibit better leadership in terms of delegating and managing their work in such a manner that they can achieve better work life balance.

The data gives a very vivid picture of how the dependent variables are associated with the independent variables in the current study. Clarity in terms of the interrelatedness of variables is also got from the data set.

Further, the data was also exposed to *f*-test for analysis of variance and the results on the *f*-test were further exposed to post-hoc test to determine the comparison within variables. This led to acceptance of two hypotheses (H_1 & H_2) while non-acceptance of H_3

It has been proven from the obtained data of the present study that income really plays a greater role in motivating the women entrepreneurs into working harder towards their business thereby helping them succeed. The result from the Chi-Square Value for association is 20.699[a] and the Significant p-Value obtained is 0.023. **Hence H_1 is accepted.**

It has been proven from the data that leadership is highly significant with experience of the respondents of running their own enterprises within the purview of this study. The correlation between Experience, an Independent Variable and Leadership, a Dependent Variable reveals the Chi-Square Value as 8.627[a] and the obtained Significant p-Value as 0.035. Leadership is also dependent upon Income, Occupation of the husband as well as partially with Number of Family Members of the respondents. There is greater significance between Number of Years of Experience of running own enterprise and Achievement Motivation, Leadership and Work life Balance (p-values=0.001, 0.002 and 0.014 respectively)

By applying Post-Hoc Studies using the Scheffe's method it is found that Achievement Motivation is highly significant with respondents having 6-10 years of experience with respondents having 2-5 years of experience (p-value=0.029) as well as those having more than 10 years of experience (p-value=0.002). **Therefore it can be established that H_2 is accepted.**

From the current study, it can be inferred that the women entrepreneurs have a balanced work life and family life. On correlating Work life Balance with the Independent variables, it was found that there is significant association between Age, Experience and Income and Work life

Balance (p-values = 0.013, 0.035 and 0.000 respectively). Even on f-Test, Work life balance is significant with experience (p-value = 0.014) thereby signifying that with more number of years in their own business, women adapt to achieving more Work life Balance; it is also significantly correlated with income, whether it is their own or that of their husband and/ or of the family (p-values = 0.000, 0.008 and 0.001 respectively). From these significant findings it can be concluded that women entrepreneurs do strike a balance between their work life and their family life and that domestic responsibilities do not hamper their entrepreneurial performance and vice versa. **Therefore, H_3 is not accepted.**

A strong conclusion which can be drawn from the quantitative data is that each entrepreneur has their own set of skills and struggles and it takes a great deal of courage to continue working despite any fruitful result visible. A lot of decisions taken while enterprising may not be immediately rewarding but they gradually enhance the capacity of the women to appreciate their own working style and their own actions.

Conclusion

From the above data, it can be concluded that Women Entrepreneurs really have made a big contribution in the nation's economy. They have taken a road less travelled to be able to achieve their goals and aspirations. They have worked tirelessly and relentlessly to achieve their targets despite the society thinking of them as a weaker sex. For them, their gender has never been a matter of shame and their aspirations have not been put down because of their gender. On the contrary, being females, they have received more recognition for being able to run successful businesses besides managing home and family.

As far as **Achievement Motivation** is concerned, Respondents enjoy reading biographies of successful people in order to derive inspiration from their lives thereby making their own work life better. Respondents plan well in time and productively utilize their free time by efficient planning. There is a great sense of achievement in women, and on knowing that people like them have achieved much in life, they are also motivated to work harder. This also hints at the aspiring and hardworking nature of the women respondents. Majority of the women respondents completely agree that they never leave a task unfinished. They continue persevering hard in order to get their tasks done. Women respondents have this innate urge to do something others have not done before an hence, they always compete with themselves to become better at their business. For them there is no substitute to hard work and they give lesser importance to luck. Moreover, they don't wait for people's approval on matters related to their work; if they feel things will work in their favour, they go ahead with their plans. Respondents agreed to befriending the most intelligent people in order to live up to the standards required in business. Respondents own their actions and believe that they themselves like to deal with difficult situations so that the blame or praise comes to them alone.

On the **Leadership** front, respondents have come across to be encouraging to their team members and have tried to foster an environment where even their subordinates learn and take up initiatives. This data also shows their interest in coaching people to be like them. The respondents are satisfied with their team members doing well and not intimidated by them. Nearly 72% of the women accept that they closely monitor the tasks at hand. A fair amount of women respondents i.e. 63% respondents believe in understanding and respecting other peoples' boundaries and limitations thereby knowing how to delegate tasks to them. This also makes them efficient leaders as they understand and accept their team members as they are. When correcting mistakes, 51% respondents do not worry about jeopardizing relationships sometimes. They feel they need to be more connected with their staff and hence may overlook mistakes however, if the mistakes can lead to serious consequences, they do put their personal relations apart from their professional ones. Since most of the respondents are self-made entrepreneurs, they have devised their own ways of leading their team. However, 48% of respondents enjoy reading articles, books and trade journals about their profession, and then implementing the new procedures which they have learned by reading.

As far as **Work-life Balance** is concerned, the respondents felt that their personal lives were not affected by their work and they could easily balance it out. Nonetheless, only 35% respondents are able to meet prescribed deadlines and schedules, without affecting their home life. 35% respondents work for extra hours to get their work done and almost 20% of the respondents often carry the additional work home. Nearly 15% respondents ignore their health and do not engage into any kind of exercises to take care of their health. However, 72% respondents admit that they thoroughly enjoy doing their work. Respondents expressed that they felt a deep sense of satisfaction by contributing to their family and the society at large by being in business and this feeling motivated them to push harder to achieve their goals.

Narratives from Selected Respondents:

During the early months of 2018, the respondent was muddling with the data collection after making all possible efforts to contact the right kind of respondents, the researcher reached out to 158 respondents across the four cities of Ahmedabad, Rajkot Surat and Vadodara. The city-wise break up of respondents is as under:

S. No.	City	No. of Respondents
1.	Ahmedabad	50
2.	Rajkot	19
3.	Surat	50
4.	Vadodara	39
TOTAL		**158**

Each respondent had a unique experience to share and almost everyone had a journey full of struggles and setbacks. But what made them what they are today is their perseverance and patience and the undying will to never give up. The researcher has tried to incorporate few narratives from the respondents as part of the study to show how women work in different trades and enterprises. The researcher has taken care of the ethical aspects and considerations before producing the case studies. Owing to these ethical considerations, the names of the respondents have not been disclosed for the purpose of confidentiality. Thus, narratives have been framed to display the issues and challenges faced by the respondents. The researcher has also attempted at trying to correlate the observations and notes along with the Social Work Skills to identify how these skills have been used by the respondents, and to what extent. Despite the pool of data available, the researcher has limited to sharing one case study per zone to justify the collected data as well as to show variety in the approach of respondents within the same state.

Making the best out of waste

1. **Name of the Organization:** WastePro, Ahmedabad
2. **Year of Establishment:** 2011
3. **Objective:** Understand, Update, Upgrade and undertake all steps possible in biodegradable waste treatment and management.
4. **Activities:** The WastePro team is a team of spirited individuals committed to change the way citizens view waste. To ensure their objectives are met, the team empower children and citizens to become change agents themselves. They also bring the Do It Yourself (DIY) concept in Waste Engagement. Moreover, by this, contribute significantly to Swachh Bharat Mission.
5. **Entrepreneurial Journey:** "I was fed up of the inaction of the municipal corporation & watching cows feeding on plastic waste in their search for food. That is when it first struck me to begin a movement to address this menace and turn into a social entrepreneur," says the respondent.

   The respondent founded her for-profit enterprise in the year 2011, after a career as an Employee for 9 years. During her employment tenure, she worked for credible organizations like ITC Ltd, Confederation of Indian Industry (CII), Autonomous University of Madrid, Spain, Mission Mangalam of the Commissionerate of Rural Development, Govt. of Gujarat and lastly the Gujarat Chamber of Commerce & Industry. The burning desire to turn an employer and start her company led the respondent to tread the uncharted path of a new venture creation specialising in Solid Waste Management. Rearing to venture a male-dominated and government-centric sector was not an easy decision for her as a woman entrepreneur. But, today as an official stakeholder, she has treated tonnes of waste and prevented waste.

   The business model of the respondent is that of treating waste in a decentralized manner, at the point of generation itself. As the project treats a unit's waste in their own premises, garden, food & paper waste is prevented from reaching landfill sites. The venture promotes decentralization of solid waste treatment and management. The enterprise deals with the sales of waste handling tools, vermicomposting tools, earthworms and bio culture.

6. **Beneficiaries:** The respondent has taken her ideas to generate revenues by selling Bioculture, Kitchen waste composting kits, selling earthworms & selling vermi-compost to bungalows, farmhouses, industries & organic farmers. Karnavati Dental School, Mayur Greens Farm House, Shri Uma Industries, Naroda and J K Lakshmi Cement Ltd., Kalol are few of the many beneficiaries.

7. **Issues and Challenges:** The respondent went home to home teaching residents to handle their waste. She has not only toiled to make Ahmedabad a greener city but she has also taught people to derive wealth from waste. She has enlightened many school going children on environment conservation and has transformed lives of many people.

   The biggest challenge that the respondent and her enterprise face today is shifting human consciousness. She believes that it is not the planet that needs saving, it is people. She believes if people start taking environment for granted, there will soon be a time when human race might just be wiped off from the face of the Earth.

8. **Correlation with Dependent Variables:** The respondent's future aspiration is to be known as one of the key social venture companies instrumental in laying the foundation for a Zero Waste Amdavad by 2030.

   The respondent works individually managing her finances and operations with quite a balance. She has a team of workers and labourers not co-workers. She finds it difficult to balance her work-life and personal life; hence, she chooses to live a life by integrating work and family thereby giving her a flexible social life.

9. **Social Work Skills Used (as perceived by the researcher)**: The researcher observed the skills of information gathering and analysis in the respondent. The researcher also observed that the respondent was assertive and direct in her approach. She considers herself as a one-person army and prefers working solo. She is good at problem solving. She has good skills of delegating the work and is inspirational to people to become more environment-conscious.

10. **Learnings:** The respondent prefers to be called a Social entrepreneur. She decided to quit her well-paying job and invest all her savings as seed money for her venture. She believes that nature does not need human beings rather human beings need nature. She has won the best stall award at Vibrant Gujarat 2013. She was a part of the stakeholder's meeting in drafting the new solid waste management rules 2013 of the Ahmedabad Municipal Corporation.

Create your own destiny!

1. **Name of the Organization:** Krish Consultancy, Rajkot
2. **Year of Establishment:** 2006
3. **Objective:** To be the Most Preferred Training and Placement Provider at India Level.
4. **Activities:** The enterprise began with two services - Job Placement (Recruitment and Staffing Solutions) and .HR & Training Implementation Services for Industries and Corporate Houses.
5. **Entrepreneurial Journey:** After Graduating, the respondent decided to start Placement Consultancy firm as she observed that people are willing to work but they do not get the chance or do not have required skills and appropriate information. So in 2006, the respondent started her own consultancy which has now grown to 25 associates across the state, 100 clients and a data base of 1 lac candidates. "Women are generally perceived as home makers with little to do with economy or commerce. However, this picture is changing. In Modern India, more and more women are taking up entrepreneurial activity especially in medium and small scale enterprises", says the respondent. She provides customized Human Resource Management and Training solutions through project based arrangements or complete outsourcing of the human resource function. She has vast experience to identify capability & skill of individual to develop their job career. She is also associated with various Management Institutes and Schools as a Resource Person and Consultant.

   She is passionate to provide best job and career counselling facilities to every job seeker of Saurashtra region through her company. She has served more than 100 Industries & corporate as a HR & Training Consultant. She is a youth icon for youngsters of Saurashtra region, especially the females.

6. **Beneficiaries:** The consultancy has grown over the years and it has been pivotal in providing jobs to over 1 lac candidates within the Saurashtra region. She has provided trainings in almost 100 industries and corporate houses. She has trained more than 10,000 Executives/supervisors in different industries & more than 50,000 students of Institutions.
7. **Issues and Challenges:** She feels that in the region where she belongs, the parents' goal is to educate the females to ensure they get the right match. This mind set hinders many women who have entrepreneurial ideas. Thus, the region sees less contribution of women in entrepreneurial activities as compared to cities like Ahmedabad and Surat.

8. **Correlation with Dependent Variables:** The respondent has contributed in growth & development of SMEs in last 13 years as 5S Expert. She is self-motivated and found space for

her venture because there was a need for such establishment in the market. She is a corporate trainer and leads others be example. She has a dedicated team of trainers and placement providers who work in cohesion to provide employment opportunities to the unemployed youth. Her enterprise follows the model of providing skill-based jobs. Since the enterprise has been long establishes, the respondent finds it very easy to strike balance between her work life and her family life. .

9. **Social Work Skills Used (as perceived by the researcher)**: The researcher observed that the respondent possessed excellent skills of communication and presentation. She possesses right knowledge and owes to her team to keep her updated on the latest news and trends that help her in her training sessions. She is resourceful and has a well-knit network of professionals who push her to fame. She possesses great leadership skills and treats her workers as family. She is empathetic and assertive. She does not believe in jeopardising personal relations at the cost if professional negligence and hence takes utmost care in dealing with her employees. She possess good interpersonal skills and problem-solving skills. She maintains a calm and pleasant disposition.

10. **Learnings:** The respondent believes that life is all about creating yourself so one must design his or her own destiny. The researcher learned that it takes years of struggle to be able to create a space for oneself in the market. The researcher observed this undying spirit to succeed at all costs, in the respondent.

Inspiring people to live WAKEFULLY!

1. **Name of the Organization:** Happy Minds, Surat
2. **Year of Establishment: 2012**
3. **Objective:** The sole mission and objective of the company is to transform lives.
4. **Activities:** HappyMinds is a multi – faceted organization that is committed to bringing this change by empowering people coming from different walks of life. HappyMinds deals in following three verticals:

   ProMinda: These Training and Workshops aim at mentoring & training individuals & organizations to enable them to reach a higher level of thinking thereby giving a meaning to their existence.

   Leadership AcademyThe Leadership Course will take you from existence to excellence! To create an impact and change the approach from traditional bookish learning to global age practical learning.

   Social Initiative-Jazba: This vertical is driven by the Jazbaa to make a difference to society in unique ways. Sessions on iLearn to iLead, Nurturing Young Trainers, Positivity Drive, etc. are the core fundamentals of this vertical.
5. **Entrepreneurial Journey:** The respondent pursued her Master's Degree in Business Administration from MIT, Pune. There she chose a module which, allowed her to work while studying and got herself placed in a Corporate training company. Her work profile was in the marketing sector but she stretched her working hours to learn as much as possible. She spent her break-times in seminars, where one of her senior used to deliver sessions on 'Life Skills'. She happily served tea and snacks to the participants just to be present there and simultaneously, learn. Even after, tiring working hours, she used to manage her time and make modules from what she learnt in the whole day. Since the beginning, she has followed the mantra – "Wake up, Dress up and Show up". And keeping the same mantra as her motivation, she soon started off as a trainer; however, small the number of audience, she had her mother and her best friend, as her constant. There were times, when people tried to pull her down. They mocked her as some 'modern-day-baba' and went on to call her a 'fanatic'. But she had a clear positive vision and a huge faith in her own work. She had built a strong inner system, relying on which she kept going on and did not let criticism disturb her pace. She then studied NLP (Neuro-Linguistic Program) where the roots of her ideology were nurtured. She then invested herself completely in understanding the thought patterns of the human mind.

6. **Beneficiaries:** HappyMinds offers services like training, seminars and workshop sessions and events on positive parenting, entrepreneurial empowerment, confidence building, communication, homemakers' empowerment, corporate training, youth conferences, positive thinking sessions and many more. The beneficiaries range from the ages of 4+ to 60+ years.
7. **Issues and Challenges:** From going door to door to share her ideas, to giving free seminars and training, the respondent dedicated her initial entrepreneurial years in spreading awareness on mind power and building a strong base of trust and understanding among her growing audience. After almost, half-a-decade of persistent efforts and hard work, she brought the wave of change she always intended to. Now, when she announces her public seminars on either 'Positive Thinking', 'Happiness Therapy', 'Mindpower' or any other Life skill subjects.

   The respondent believes that even today, there is not much awareness on the importance of mental wellbeing. She believes a lot needs to be done and she needs to expand the horizons of her venture to different parts of the country and the world at large.
8. **Correlation with Dependent Variables:** The respondent at the age of 19, dreamt and planned out a certain kind of life for herself and did every possible thing to make it work. However, fate had other plans altogether, none of it worked out the way she wanted. Broken and shattered, she started once again. This time her mind was railing with the idea of understanding how it all works. She intended to know how the mind over matter changed the game. That is precisely the time when things started playing out in her favour. This shows the level of motivation of the respondent. She is highly motivated by her own success and aspires to do more to make the world a better place. From a trainer, she has now become an influencer.

   The organisation has a flat model, having more than 25 volunteers associated with them. The respondent, who is the boss, chooses to be called Director and Chief Mentor while she prefers to call her staff members as Chief Energy Officers (CEOs). She takes fun seriously and ensures all employees are always happy and cheerful. She follows a leadership model that further empowers the staff and volunteers to stay committed to the cause of transformation.

   At an early age of 24 years, the respondent was the youngest speaker on the panel across the country for her Social Contribution at the TEDx TALK platform. BRITT India and iHeart Global Inspiration Women'17 have honoured her as one of the top 60 women entrepreneurs in India with innovative business ideas in 2016.

   The respondent owes her success to her extremely supportive mother, her loving husband and both her families. Moreover, the positivity that her professional life gives her makes her personal life bloom too. Hence, she does not face any significant work-life balance issues.

9. **Social Work Skills Used (as perceived by the researcher)**: The respondent is an extremely confident individual with a very clear understanding of herself and her work. She lives a purpose-driven life and chooses to transform people's lives through her efforts. From the perspective of social work skills, the researcher has observed that the respondent possess excellent skills of communication and presentation. She has great resource mobilization skills and very sound interpersonal as well as intrapersonal skills. She follows the POSDCORB model thoroughly and is an efficient manager. She possesses good problem solving skills. She is accepting of herself and the people around her and believes in holding a non-judgmental attitude towards others. She is an epitome of positivity, enthusiasm and energy.
10. **Learnings:** The researcher learnt about a new area of entrepreneurship, which dealt with infusing leadership among children as young as 4 years. On her visit to the enterprise, the researcher observed how intricately confidence boosting and leadership training is provided to the young minds. The researcher also observed how patiently and calmly the staff engaged with children in their fun activities. It was surprising to see that many parents are actually interested in the leadership camps for children and are willing to pay any amount just to enrol into the HappyMinds Academy.

Serving bowlful of happiness

1. **Name of the Organization:** Messy Bowl (earlier Maggewala), Vadodara
2. **Year of Establishment:** 2013
3. **Objective:** To pioneer in delivering hand-crafted food to people and savor their taste buds.
4. **Activities:** The USP is to create usual instant noodles with a twist and offer variety of flavours for all age groups. Their dishes are served at competitive and affordable prices. Core reason of our good standing in the market is our quality that has authentic flavours. It helps them become honest & trust-worthy brand among Maggi-lovers.
5. **Entrepreneurial Journey:** The respondent ventured into food industry without any prior experience. The respondent was working at the Vadodara Municipal Corporation making a sound living with a regular paycheque rolling in every month. Her life was near to perfect. She was also married to the person of her choice, which made her feel that her life was complete. However, life had a different take on her perceived happiness. She soon realised she was in to a troubled marriage and wanted to help her husband by running her venture. However, her husband belonged to a family conditioned to believe that a stable job is what one needs and abhorred the aspect of entrepreneurship. Moreover, the husband was not happy in letting his wife work while he had no income. However, the respondent stood against all odds to make sure her venture succeeded. Her family and friends immensely supported her and she truly believes that her legal separation from her life partner was like receiving a new life. She is an example of what a woman, who has only self-belief by her side, can do. With her vision to serve best in class quality food at affordable prices, she started a brand, that is today, loved by people of all age groups.
6. **Beneficiaries:** From one outlet in the heart of the cultural capital of Gujarat (Vadodara), it became one of the fastest growing chain in the fast-food industry.
   She has created a pool of happy customers in various parts of the country. Besides having their outlets in Vadodara, the venture has also expanded into cities of Rajkot, Surat, Hyderabad and Chennai.
7. **Issues and Challenges:** Founded in 2013 by a first generation entrepreneur with a belief in pioneering in delivering handcrafted food to people and savour the taste buds. Core reason of their strong foothold in the market, despite selling varieties of only one product, is their quality that has authentic flavours. This strategy helps them become honest & trust-worthy brand among Maggi-lovers. "We do not have a fancy, elaborate menu but whatever we serve is cooked with all our heart", says the respondent.

They faced a big challenge when, in 2015, the Nestle Group landed under a legal scanner being accused of being unsafe and hazardous for consumption. As against the maximum limit of 2.50 parts per million (ppm), the amount of lead detected in the Maggi samples was perilously high at 17.2 ppm. The second offence was mentioning 'No added MSG' on the packaging, which is an act of mislabelling. On June 5, 2015, a nationwide ban was imposed on Maggi by FSSAI (Food Safety and Standards Authority of India). During this phase, the respondent had to go out of business for a while s there was a public outrage on serving a banned product. The respondent had to change the menu at her restaurant chain and had to wait until the matter cleared. That was a time when the respondent's outlet was named Maggewala. Once Nestle re-established itself back in the market, the respondent changed the name of her outlet from Maggewala to Messy Bowl. She continues to serve the same product with guaranteed safety. The product and the food chain gained such popularity that she was invited as a guest at the TV Show - Maggi Kitchen Journeys, hosted by the renowned Bollywood actress Smt. Renuka Shahane Rana (

8. **Correlation with Dependent Variables:**The respondent kept a very low profile and was happy with her life. After her marriage and shift to Ahmedabad, she came across many street vendors who were experimenting and serving Maggi noodles. On one such evening, when she was out with her friends sharing a bowl of Maggi, it occurred to her to leverage on this idea and open a chain in her hometown – Vadodara. Things seemed to be falling in place, until she found out that her husband did not have a job and she stepped forward to help with her venture. She was shocked to learn that the husband was not in favour of her working and restricted her movement. It was then that her mother stepped forward and helped her come out of a failed marriage. Leaving this behind, the respondent felt more confident in taking the leap and was motivated to devote more hours to her venture.

The respondent was supported immensely by her family and few friends who helped her choose the right location, staff, furniture, marketing, advertising, publicity and streamlining things for her. She believes in bringing people together and leads by example. She feels if one provides proper attention to the details, it can win many hearts. She does not believe in authority and treats her staff as family.

Since initially the venture required more involvement, the respondent had to spend days and nights at the restaurant ensuring that everything was in order. At that time, it was difficult to manage her work life and her family life. Gradually, when things started working out favourably, she was able to take time off for her personal commitments. Today, she has a well-balanced life at both personal and professional fronts.

9. **Social Work Skills Used (as perceived by the researcher)**: The researcher observed that the respondent possessed great managerial skills. She had a vision and was able to persevere through all the troubles to make sure her vision turned to reality. She applies the POSDCORB model very efficiently. She displays excellent interpersonal and intrapersonal skills. She possesses sound leadership skills and good skills of resource mobilization. She thrives on her networking skills, which have led to the nationwide expansion of her outlets. She has great problem solving skills. She maintains a pleasant disposition and is always ready to continue to inspire young and potential entrepreneurs with her self-belief and positivity.
10. **Learnings:** The researcher learned that eureka moment could also occur during the times one least expects it. In case of the respondent, it occurred over a bowl of Maggi. Today, that has turned into a successful business venture. The researcher learned that entrepreneurial success does not come overnight and entrepreneurs have to persevere hard to ensure that their vision becomes a reality. The researcher also learnt that despite the difficulties, one needs to pursue and put their foot down to create great results.

SUGGESTIONS & RECOMMENDATIONS

One of the objectives of the study was also to suggest a framework to promote women entrepreneurship in Gujarat. A framework for promoting entrepreneurship among women in Gujarat will be duly discussed and suggested in the next chapter.

This chapter presents the suggestions of the researcher and the recommendations along with areas of Social Work Intervention based on the findings of the analysis elaborated in the previous chapters.

This chapter is further divided in to two parts:

1. Suggestions & Recommendations

2. Areas of Social Work Interventions

1. Suggestions and Recommendations

One of the objectives for this study was to also provide a framework to develop entrepreneurship among women in Gujarat. From the study undertaken, the researcher has got many new insights about female entrepreneurship which were not known earlier. However, there are certain loopholes, to cover up which, the researcher provides suggestions as part of the study.

It is said that an entrepreneur is someone who possesses a risk taking ability and the courage to face problems and suggest alternate and novel suggestions for the same. In case of women entrepreneurs, the expectation is different because if their risk fails, they have no other choice but to shun their dreams of enterprising. It is with this perspective that the suggestions have been provided.

a. Promoting positive attitude towards entrepreneurship through role models

One of the biggest reasons women remain laid back in the field of entrepreneurship is the negative conditioning towards failure and a deep understanding that risk taking almost always leads to failure. This negativity can only be countered with a positive infusion of successful women entrepreneurs taking the lead in coaching the nascent women entrepreneurs.

- Enterprising women are different from small scale vendors and hence their psychosocial outlook is very different. Their education and skills can be effectively used for creating proper forums for budding entrepreneurs wherein the stable entrepreneurs can explain the challenges and mentor the budding entrepreneurs to nurture better.
- The dedicated forums for women entrepreneurs can offer periodical trainings which can cater and invite more and more participation from enterprising women in order to hone their entrepreneurial skills.
- Voluntary role-model engagement programmes can be encouraged where successful women entrepreneurs come together at public events to encourage entrepreneurship among women.
- These forums can be utilised to incubate budding entrepreneurs wherein the skill development can be rigorously emphasised.
- These forums can also act as a platform for women entrepreneurs to effectively conceptualise their entrepreneurial ideas and convert it in to successful businesses.

b. Approaching the proper authorities

- There is an increase in the emphasis of promoting women entrepreneurship under the aegis of Skill India Program of the Government of India and the state government as well. However, there are many entry points for women and many times this creates a confusion among those who are actually seeking help.
- It is suggested to have a single-window system for the registration to loans to scholarships and benefitting schemes, and other aspects related to entrepreneurship. Here, it becomes important to understand that women entrepreneurs may not always have a self-help group (SHGs) or support groups, they may be individuals willing to start up a firm on their own. For such women, a separate system must be devised without them having to run from pillar to post to ensure their paperwork is in place.
- Use of the internet and online systems can prove to be a great boon to the government functionaries and the beneficiaries alike in order to streamline processes for every aspect of entrepreneurship.

- Enterprises registered by women in the name of women must be specially recognised and some rewards can be provided to those enterprises. This strategy may encourage other women also to come up with novel ideas and can lead to increased entrepreneurship.

c. Facilitate access to finance through financial literacy and a range of financing instruments

- There are various financial schemes available for women entrepreneurs in India but the beneficiaries are unaware about it because e of lack of awareness and proper advertisement.
- It is observed that women entrepreneurs lack financial literacy and that financial institutions are also not very open about the explanation regarding the complicated policies.
- Modern methods like "fintech" as an industry sector is changing the nature of business funding, and small businesses are increasingly using a wider range of sources of capital.
- Use of Crowdfunding is also becoming more popular and is applied in many forms, including rewards or gifts, debt, equity, or a combination of debt and equity. Crowdfunding assures of reaching out to a large network of potential financiers over the internet without any face-to-face contact. By resorting to these modern methods, female entrepreneurs may be able to overcome the traditional discrimination in accessing financial backing to start a new ventures. (Halabisky, Potter, Garcia, & Levie, 2016)

d. Easing journey from the mind to the market

- Entrepreneurship is all about the right ideas at the right time combined with the right resources and we have a successful enterprise. However, when it comes to enterprising women, a lot of ideation remains limited only within the enclosures of their minds. The scepticism of failing does not let women take many risks. Moreover, the linkages between the production and marketing of the same are found to be very weak and in their nascent stage.
- Inability to promote their own products or services leaves women dejected and withdrawn from entrepreneurship. This can be explained as every woman knows how to cook, but not every woman is a restaurateur.
- In Gujarat, a government arm of the Mission Mangalam - Gujarat Livelihoods Promotion Company (GLPC) Limited has initiated exhibitions where women entrepreneurs can rent out stalls to market their products. However, these kind of initiatives need to be increased and promoted at a higher level so as to cater to a larger population.

e. Promoting Work life Balance and Family Counselling

- The social roles provoke women to ensure the house is kept in proper order while the man is at work. However, this cannot be convincingly said in case if the women begin working and leave the house for the significant other members to take care of. Hence, sensitization within families needs to be created so that they can promote women in their entrepreneurial activities.
- Young female entrepreneurs also have the responsibilities of marriage and family due to which their entrepreneurial role takes a back seat. A supportive and understanding family can help women overcome their inhibition even during these life events.
- Sensitization to explain that women can also portray the role of an entrepreneur must be created right from the beginning.
- Schools and colleges must adopt Entrepreneurship as part of their curriculum so that idea generation can start at a very young age irrespective of the gender and that deeper family understanding is created to ensure proper support.

2. Areas of Social Work Interventions

The accepted Global definition of Social Work Profession defines that "Social work is a practice-based profession and an academic discipline that promotes social change and development, social cohesion, and the empowerment and liberation of people. Principles of social justice, human rights, collective responsibility and respect for diversities are central to social work. Underpinned by theories of social work, social sciences, humanities and indigenous knowledges, social work engages people and structures to address life challenges and enhance wellbeing." (International Federation of Social Workers, 2014)

Hence, the professional social work respects diversity and enhances well being by use of various methods by which an individual, group or community at large can be empowered to help itself even in absence of any individual or situation.

The gamut of social work has always focused its attention on the weaker and vulnerable sections of the society and tried to mainstream them by countering the difficulties and enhancing welfare.

Through the current study also, the researcher has tried to bring about interventions which can augment entrepreneurship among women and help them be self-reliant.

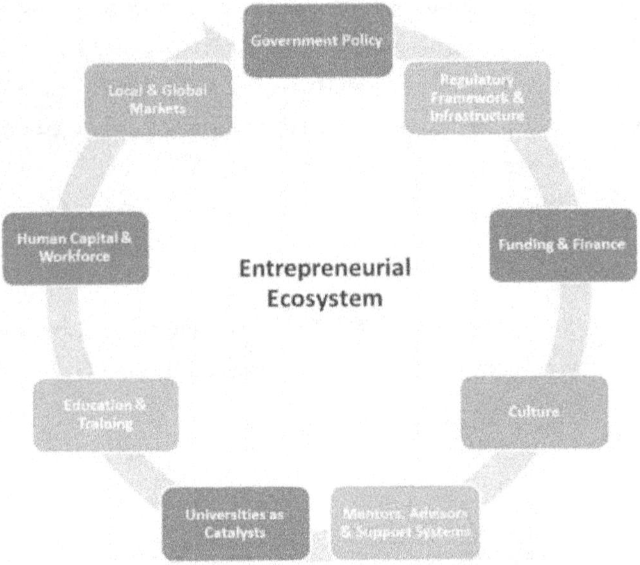

Figure 6.2.1 Showing Entrepreneurial Ecosystem

Figure 6.2.1 briefly explains the entrepreneurial ecosystem in which there are a lot of factors which play their role on helping the entrepreneur rise above their inhibitions and become a successful business-owner. Each of these elements of the entrepreneurial ecosystem can be linked with Social Work Methods to prepare a proper framework for intervention.

Social Work has two methods:

1. Primary Methods
    1.1. Social Casework
    1.2. Social Groupwork
    1.3. Community Organization
2. Secondary Methods
    2.1 Social Welfare Administration
    2.2 Social Work Research
    2.3 Social Action

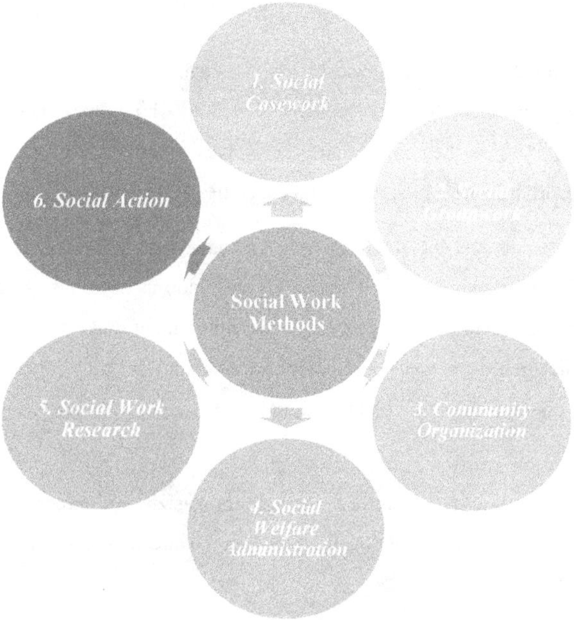

Each method is unique in its approach to ease out human lives by providing scope for improvement at every aspect. Each aspect is described in detail below with areas for Social Work intervention at various levels to augment and accelerate the growth of women entrepreneurship in the state and the country at large.

1. Social Casework

Mary Richmond (1917) defined Social Casework as "the art of bringing better adjustment in social relationships of men, women or children." The purview of casework is dealing with individual problems and help them overcome it.

By applying the theories of casework to the current study, women can be empowered with greater confidence within themselves. In the Indian context, family dependency and personal obligations on women create hindrances to them in succeeding in their career. Motivational factors: Factors like intrinsic motivation, government policies, financial assistance and family support are still in weak shape in India for women entrepreneurs. These individual issues can be taken up and resolved in order to facilitate women. Providing proper hand-holding until these women are fully confident in the way they carry themselves in the market.

2. Social Groupwork

Trecker (1955) defined Social Groupwork as "a method through which individuals in groups and in agency settings are helped by a worker who guides their interaction in programme activities so that they may relate themselves to others and experience growth opportunities in accordance with their needs and capacities to the end.

With the help of the method of Social Groupwork, a focused approach can be followed to work in the direction of empowering women. Continuous and rigorous trainings can be provided to a small group of budding women entrepreneurs and helping them understand group processes and ways to handle their employees and practice leadership. With constant and consistent efforts in training and skilling the women entrepreneurs, they can be made to become more confident as a group.

Moreover, regular meetings with groups of women emphasizing on experience sharing and testimonies can help other women gain inspiration from the already successful women.

3. Community Organization

Ross (1967) defines community organisation as a "process by which a community identifies its needs or objectives, gives priority to them, develops confidence and will to work at them, finds resources (internal and external) to deal with them, and in doing so, extends and develops cooperative and collaborative attitudes and practices in the community".

A reference-network of women entrepreneurs also can be considered as one community from the many communities of the society. This community can be made self-sustaining by developing confidence in them, creating awareness of the various schemes which can provide financial assistance and organizing training programs for them.

Moreover, the various methods of community organization like Education promotion to promote entrepreneurial education, Legislative promotion to promote various schemes offered by the government organizations, non-legislative promotion to promote the schemes offered by voluntary organizations, fund raising for enterprises, conferences of reference networks and evaluation of the status of enterprises can also be done for the smoot functioning of enterprises.

4. Social Welfare Administration

Harleigh Tracker (1971) interprets social welfare administration as a "process of working with people in ways that release and relate their energies so that they use all available resources to accomplish the purpose of providing needed community services and programmes." (Devi & Ravi, 1998)

The government agencies and private agencies which exist in providing support to the enterprising women can be made more active in their functions so that a regular follow-up on trainings is taken and need-based programmes are designed for the benefit of the women entrepreneurs. The District Industries Centers (DICs), the Centres for Entrepreneurship Development, Chambers of Commerce and Industries, national and cooperative banks and all other significant agencies should work in collaboration in order to properly assist women in their trade and business guidance. A single-window agency system must be implemented wherein women can get their issued resolved under one single roof.

5. Social Work Research

Social work research may be defined as systematic investigation into the problems in the field of social work. (Das, 2000)

There is a great scope of researching in the field of women entrepreneurship from a social work perspective. The researcher has made a humble attempt in identifying and studying certain psychosocial issues of entrepreneurship among women but many areas such as impact of government schemes, impact of financial literacy, market linkages and women, leadership, gender analysis of entrepreneurship can be undertaken to gather insights from the field and can be researched upon.

Research in any topic will yield to generation of further scope of study. Similarly, the current study though has limitations, yet many researches can be formulated on the lines of the same topic as that under current study.

6. Social Action

Singh (1986) explains that social action is a process in which conscious, systematic and organised efforts are made by some elites and/or people themselves to bring about change in the system which is instrumental in solving problems and improving conditions which limit the social functioning of weaker and vulnerable sections. (Mishra P. D., 1992)

It is only by collaborating the various efforts made by the various agencies that something positive in the direction of helping women entrepreneurs grow and develop well can be achieved. Proper use of advocacy of the rights of women and women entrepreneurs can help foster the area of entrepreneurship among women. Models of Locality Development and Social Planning can be effective in practicing Social Action within the entrepreneur community.

Thus by working on these aspects, a proper intervention strategy for empowering women through entrepreneurship can be made.

From the above, it can be summarized diagrammatically that development of women entrepreneurship basically requires certain common and collective actions to be undertaken. If designed and implemented well, these actions can pave a new road towards achieving and equal society for women entrepreneurs. These actions are:

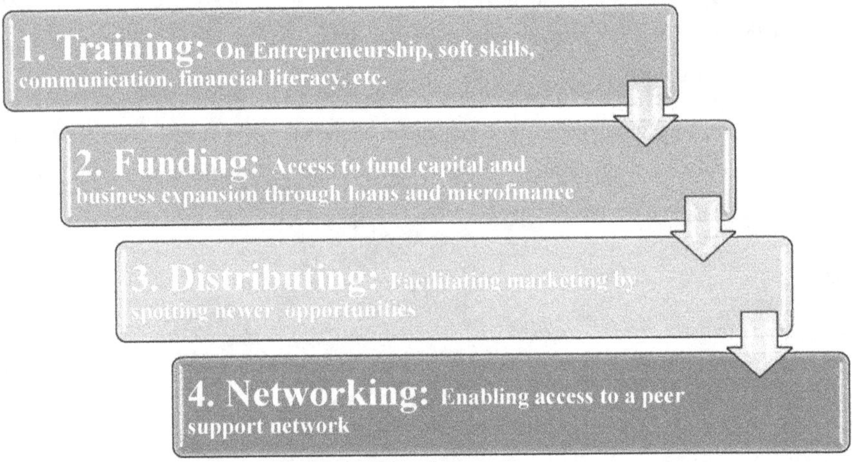

1. Training: Effective trainings can be planned and arranged for focused groups of women entrepreneurs wherein they can be trained on various entrepreneurial aspects like Proposal Formation, Soft Skills, Communication, Financial Literacy, Banking Procedures, Government

Schemes available for support, collaborating and networking as well as overall entrepreneurship development. A proper handholding mechanism must be designed which helps women overcome their diffidence.

2. Funding: Proper linkages must be created between financing agencies and the beneficiaries. Technology and crowdfunding can also be used to its fullest potential to encourage women entrepreneurs. Understanding the need of the business and learning to respond to it by expanding the capital investment with the help of loans and other means of financial support must be encouraged among women. Awareness of these facilities must be spread through proper media so that the number of beneficiaries may rise.

3. Distributing: Effective market linkages must be created in order to facilitate and encourage greater participation from women entrepreneurs. Due to lack of avenues to market their products and services, many women give up on their entrepreneurial dream. This can be avoided if a facilitating market connect can be established with the help of appropriate authorities. Internet has revolutionised the way businesses work these days. Understanding the using digital platforms and social media for marketing can be introduced so that women entrepreneurs can easily market their products online.

4. Networking: Businesses are known to flourish if one has the right kind of contacts. Hence networking becomes a very important aspect of developing entrepreneurship. Dedicated platforms and forums can be created where budding women entrepreneurs can easily approach and get proper guidance, advice and mentoring on how to grow and expand their business. This can enable a learning ecosystem for the beneficiaries and every individual, experienced or inexperienced in managing their business will be benefitted thereby creating a win-win situation for all.

As the society is undergoing a major paradigm shift towards technology, there is requirement to modify the society beliefs and provide women entrepreneurs with a strong ecosystem for entrepreneurship and democracy.

On a concluding note, the researcher would like to quote a very famous poem from one of the very well known and renowned Afro-American Women's Rights Activist, Maya Angelou (Angelou, 2011) on women. The focus and context of her work was basically expressing the extraordinary characteristics of women, however, the researcher has taken liberty to change some words in order to make it more meaningful and apt in the context of the current study.

Home makers wonder where my secret lies

Working women question my business size

But when I start to tell them,

They think I'm telling lies.

I say, it's in the reach of my arms,

the span of my waist

the stride of my step

the curl of my hair, in my fingers, interlaced

I'm a Woman. Phenomenally.

Phenomenal Woman. That's Me.

Inspired by
Maya Angelou

Bibliography

A String of Pearls - Women Entrepreneurs of Sri Lanka: A Casebook. (1995). New Delhi: Asian and Pacific Centre for Transfer of Technology.

Akhtar , S., Nagien, S., & Khan, P. (2009). *Economic Development: A Prospective Study of Women's Entrepreneurship.* (B. Kumar, Ed.) New Delhi: Regal Publications.

Allen, D., Herst, L., Bruck, S., & Sutton, M. (2000). Consequences associated with work to family conflict: A review and agenda for future research. *Journal of Occupational Health Psychology, V*(2), 278-308.

Ambedkar, B. R. (1987). Women and Counter Revolution (Riddles of Hindu Women). In B. R. Ambedkar, *Dr. Baba Saheb Ambedkar: Writings and Speeches* (Vol. III). Department of Education, Govt of Maharashtra.

Anderson, R. S., Grant, M. M., & Speck, B. W. (2008). *Technology to Teach Literacy: A Resource for K-8 Teachers.* Michigan: Pearson/Merrill Prentice Hall.

Angelou, M. (2011). *Phenomenal Woman: Four Poems Celebrating Women.* New York: Random House Publishing Group.

Arya, S. (2000). *Women Gender Equality and the State.* New Delhi: Deep and Deep Publications.

Banerjee, T. (2012, March 11). *Here's How The Status Of Women Has Changed In India [Since 1950 Till Date].* Retrieved March 29, 2017, from YouthKiAwaaz: https://www.youthkiawaaz.com/2012/03/heres-how-the-status-of-women-has-changed-in-india-since-1950-till-date/

Bangwal, D., & Tiwari, P. (2015). Green HRM – A way to greening the environment. *IOSR Journal of Business and Management (IOSR-JBM)* , 45-53.

Bansal, R. (2013). *Follow Every Rainbow.* Chennai: Westlan Ltd.

Barnwal, B. K. (2014, Mar-Apr). Dr. B. R.Ambedkar's Quest for Gender Equality It's Relevance in Contemporary Feminist Discourse. *Online International Interdisciplinary Research Journal, Volume-IV*(Issue-II).

Bent-Goodley, T. B. (2002). DEFINING AND CONCEPTUALIZING SOCIAL WORK ENTREPRENEURSHIP. *Journal of Social Work Education*, 291-302.

Bhansali, V. K. (1999). *Entrepreneurship Development in Next Century: Entrepreneuship, STrategic Managemement and Globalization.* Jaipur: Rawat Publications.

Brush, C. G. (2009, September). *Women Entrepreneurs: A Research Overview.* Retrieved from Oxford Handbooks Online:

http://www.oxfordhandbooks.com/view/10.1093/oxfordhb/9780199546992.001.0001/oxfordhb-9780199546992-e-23

Carlson, S. D., Kacmar, M. K., & Williams, J. L. (2000). Construction and Initial Validation of a Multidomensional Measure of Work-family Conflict. *Journal of Vocational Behaviour*, 249-276.

Carr, M. (1990). Women in Small Scale Industries - Some Lessons from Africa. *Small Enterprise Development, I*(I), 47.

Carsrud, A. L., & Brannback, M. E. (2007). *Entrepreneurship.* Westport, Connecticut: Greenwood Publishing Group.

Cassidy, T., & Lynn, R. (1989). A multifactorial approach to achievement motivation: the development of a comprehensive measure. *Journal of Occupational Psychology*, 301–312.

Chakraborty, D., & Barman, R. R. (2014). A Study on Impact of Motivational Factors on the Growth of Women Entrepreneurship in Assam. *IOSR Journal of Business and Management, 16*(03), 71-80. Retrieved May 2017

Charantimath, P. (2014). *Entrepreneurship Development - Small Business Entreprise.* New Delhi: Pearson Education India.

Chaturvedi, A. (2016, April). A STUDY OF GREEN HRM PRACTICES ADOPTED IN THE SELECTED INDUSTRIAL ORGANIZATIONS . Vadodara: Unpublished.

Chaturvedi, R. (2015). *Work-life Balance of Women Workers.* Vadodara: Unpublished Dissertation.

Cherunilam, F. (n.d.). *Business Environment: Text and Cases.*

Chidambaram, K., & Alagappan, V. (1999). *Business Environment.* New Delhi: Vikas Publishing House.

Chitkara, M. G. (2002). *Dr. Ambedkar and Social Justice.* New Delhi: APH Publishing.

Christensen, J. (2015). Acting locally, Thinking Globally in Social Work Education. *Social Work Education in Europe: towards 2025*, (pp. 1-15). Milan.

Clifton, D., & Amran, A. (2011). The Stakeholder Approach: A Sustainability Perspective. *Journal of Business Ethics*(98), 121–136.

Coate, S. J. (2004). *Ecology and Social Work.* New York: Paul & Co.

Cochran, T. C. (1968). Entepreneurship. In D. L. Sills, & R. K. Merton, *International Encyclopedia of the Social Sciences* (p. 87). New York: Macmillan Co. and Free Press.

Collins, C. J., Hanges, P. J., & Locke, E. A. (2004). The Relationship of Achievement Motivation to Entrepreneurial Behavior: A Meta-Analysis. *Human Performance, 17*(1), 95-117.

Collins, C. J., Hanges, P. J., & Locke, E. A. (2004). The Relationship of Achievement Motivation to Entrepreneurial Behavior: A Meta-Analysis. *Human Performance, XVII*(1), 95-117.

Coughlin, J. H., & Thomas, A. R. (2002). *The Rise of Women Entrepreneurs: People, Processes, and Global Trends.* Westport, Connecticut: Quorum Books.

Cox, D., & Pawar, M. (2006). *International Social Work: Issues, Strategies and Programs.* New Delhi: Vistaar Publications.

Crosby, B. C., & Bryson, J. M. (2005). *Leadership for the Common Good: Tackling Public Problems in a Shared-Power World.* San Francisco: Jossey-Bass.

Dale, M. (2012, January). *Social Work Takes on Business*. Retrieved from NASW NEWS: http://www.socialworkers.org/pubs/news/2012/01/social-work-business.asp

Damor, K., & Mishra, P. (2016, April 22). Entrepreneurs check into women-only industrial park. *The Times of India.*

Danto, E. A. (2011, May 25). *Postmodernism and Social Work.* Retrieved from Oxford Bibliographies: http://www.oxfordbibliographies.com/view/document/obo-9780195389678/obo-9780195389678-0132.xml

Das, D. K. (2000). *Practice of Social Research.* Jaipur: Rawat Publications.

Das, D. K. (2000). *Practice of Social Research : A Social Work Perspective.* Jaipur: Rawat Publications.

Das, R. (1998). *The Environment Divide: the Dilemma of Developing Countries.* New Delhi: Indus.

Davis, G. (2010). *Closing Key-note Address.* New York: United Nations Headquarters. Retrieved March 28, 2017

Deepika, R., & Kapargam, V. (2016). A study on green HRM practices in an organisation. *International Journal of Applied Research*, 426-429. Retrieved 2017

Desai, M. (2002). *Ideologies and Social Work: Historical and Contemporary Analyses.* Jaipur: Rawat Publications.

Desai, V. (2007). *Entrepreneurship Development and Management.* Mumbai: Himalaya Publishing House.

Desai, V. (2009). *The Dynamics of Entrepreneurial Development and Management.* New Delhi: Himalaya Publishing House.

Deshmukh, S., & Deshmukh, P. (2016). Improving HR Interventions Related to Work Life Balance for Educational Sector in Indian Scenario. *Intercontinental Journal of Human Resource Research Review*, 165-176.

Devi, R., & Ravi, P. (1998). *Social Work and Social Welfare Administration, Methods and Practices (Vol I).* Jaipur: Mangal Deep Publications.

Dhameja, S. K. (2002). *Women Entrepreneurs: Opportunities, Performance and Problems.* New Delhi: Deep & Deep Publications Pvt. Ltd.

Dhani, S. L. (2007). *Dr. B.R. Ambedkar: Man of Millennium For Social Justice.* New Delhi: Kalpaz Publications.

Dhanvijay, V. (2012, April). Dr. Babasaheb Ambedkar's Efforts for Women Empowerment and Present Status of Women in Society. *Electronic International Interdisciplinary Research Journal, Volume-I*(Issue II).

Dhawan, A., & Dalmia, N. (2015, July 08). The Digital India Push for Education. *The Financial Express*.

Dixit, M. (1998). *Women and Achievement: Dynamics of Participation and Partnership.* New Delhi: Kanishka Publishers, Distributors.

Dobers, P. (2009). Corporate Social Responsibility: Management and Methods. *Corporate Social Responsibility and Environmental Management*, 185-191.

Dodh, P., Singh, S., & Ravita. (2013). Corporate Social Responsibility and Sustainable Development in India. *Global Journal of Management and Business Studies*, 681-688.

Dominelli, L. (1990). *Women and Community Action.* Birmingham: Venture Press.

Dominelli, L. (2005). *Social Work: Theory and Practice for a Changing Profession.* Massachusetts: Polity Press.

Drucker, P. (1973). *Management: Tasks, Responsibilities, Practices.* New York: Harper & Row.

Drucker, P. (2014). *Innovation and Entrepreneurship.* Routledge: New York.

Drucker, P. F. (1986). *Innovation and Entrepreneurship, Practice and Principles.* London: Pan Books.

Dubrin, A. J. (2004). *Leadership: Research Findings, Practice and Skills* (4th ed.). New York: Houghton Mifflin Company.

Eddleston, K. A., & Powell, G. N. (2012). Nurturing Entrepreneurs' Work–Family Balance: A Gendered Perspective . *Entrepreneurship Theory and Practice*, 513-540.

Eraut, M. (1994). *Developing Professional Knowledge and Competence.* London: Falmer Press.

Erickson, C. L. (2018). *Environmental Justice as Social Work Practice.* New York: Oxford University Press.

Feminism. (2015, August 14). Retrieved from UK Essays: https://www.ukessays.com/essays/social-work/feminism.php?cref=1

Fielden, S. L., & Dawe, A. (2004). Entrepreneurship and Social Inclusion. *Women in Management Review, 19*(3), 139-142.

Finn, J. D. (2000). *Extending Education Through Technology: Selected Writings by James D. Finn on Instructional Technology.* (R. McBeth, Ed.) Connecticut: Information Age Publishing.

Flexner, S. B. (1987). *Random House Webster's Unabridged Dictionary.* London: Random House.

Fondas, N. (1997). Feminization unveiled: Management Qualities in Contemporary Writings. *Academy of Management Review*, 257-282.

From Ideas to Independence: A Century of Entrepreneurial Women. (n.d.). Retrieved from National Women's History Museum: http://entrepreneurs.nwhm.org/#/1910-1939/1

Ganesan, S., & Duraipandian, R. (1999). Promoting Entrepreneurship among Rural Women. In M. Soundarapandian, *Women Entrepreneurship: Issues and Strategies* (pp. 62-67). New Delhi: Kanishka Publishers.

Geetha, P., Prema, M., & Jansi, M. (2012). *Women in Management: Leadership and Capacity Building.* New Delhi: Mittal Publications.

Giddens, A. (2001). *The Consequences of Modernity.* London: Polity Press.

Global Definition of Social Work. (2014, August 06). Retrieved from IFSW: http://ifsw.org/policies/definition-of-social-work/

Goffee, R., & Scase, R. (1985). *Women in Charge: The Experiences of Female Entrepreneurs.* London: Unwin Hyman.

Gond, J.-P., & Moon, J. (2011). Corporate Social Responsibility in Retrospect and Prospect:Exploring the Life-Cycle of an Essentially Contested Concept. In R. Caruana, *Routledge Major Work on Corporate Social Responsibility* (pp. 01-40). Nottingham: Nottingham University Business School.

Gray, M., Coates, J., & Hetherington, T. (2012). *Environmental Social Work.* Routledge: New York.

Greaves, T. W. (2012). *Revolutionizing Education Through Technology: The Project RED Roadmap for Transformation.* Virginia: International Society for Technology in Education.

Green, E., & Cohen, L. (1995). 'Women's business': Are women entrepreneurs breaking new ground or simply balancing the demands of 'women's work' in a new way? *Journal of Gender Studies*, 297-314.

Greenhaus, H. J., & Collins, H. K. (n.d.). The Relation between Work Family Balance and Quality of Life. *Journal of Vocational Behaviour*, 510-531.

Greenhaus, H., Collins, H., & Shaw, D. (n.d.). The relation between work family balance and quality of life. *Journal of Vocational Behaviour*, 510-531.

Gretczko, M., & Mazor, A. (2016, March 10). Designing HR for the Moments that Matter. *Human Capital*. Retrieved July 2018, from https://hrtimesblog.com/2016/03/10/designing-hr-for-the-moments-that-matter/

Guenther, J. (Ed.). (2015, September 07). *Girl Child Labourers in India: An Invisible Issue.* Retrieved September 20, 2015, from WikiGender: http://www.wikigender.org/index.php/Girl_Child_Labourers_in_India:_An_Invisible_Issue

Guha, R. (2011). *Makers of Modern India.* Cambridge: Harvard University Press.

Gunjal, V. R. (2012). Dr. Babasaheb Ambedkar and Women Empowerment. *Social Work*, 84-85.

Gurnani, P. S. (2016). *Women Entrepreneurship: Emerging Dimension of Entrepreneurship in India.* New Delhi: Educreation Publishing.

Halabisky, D., Potter, J., Garcia, M. C., & Levie, J. (2016). *Policy Brief on Women's Entrepreneurship.* Luxembourg: Publications Office of the European Union. doi:10.2767/50209

Hall, J., Matos, S., Sheehan, L., & Silvestre, B. (2012). Entrepreneurship and Innovation at the Base of the Pyramid: A Recipe for Inclusive Growth or Social Exclusion? *Journal of Management Studies, 49*(4), 785–812.

Halme, M., Roome, N., & Dobers, P. (2009). Corporate responsibility: Reflections on context and. *Scandinavian Journal of Management*, 1-9.

Harari, Y. N. (2015). *Sapiens: A Brief History of Humankind.* London: Vintage-Penguin Random House.

Hegde, B. (2013). Dr. B. R. Ambedkar: A Ray of Hope for the Downtrodden in India. *International Journal of Social Science & Interdisciplinary Research*, 144-147.

Hisrich, R. (2011). *Entrepreneurship 6/E.* New Delhi: Tata McGraw-Hill.

Hisrich, R. D., Peters, M. P., & Shepherd, D. A. (2005). *Entrepreneurship.* Boston, MA: McGraw-Hill Irwin.

Holmqvist, M. (2009). Corporate social responsibility as corporate social control: The case of work-sitehealth promotion. *Scandinavian Journal of Management*, 68-72.

Hughes, K. (2006). Exploring Motivation and Success Among Canadian Women Entrepreneurs . *Journal of Small Business & Entrepreneurship*, 107-120.

IANS. (2016, May 07). Women run 14 per cent of Indian businesses, most self-financed . *The Economic Times - Panache*.

Ibarra, H., Ely, R. J., & Kolb, D. M. (2013, September). Women Rising: The Unseen Barriers. *Harvard Business Review*.

International Federation of Social Workers. (2014, July). *Global Definition of the Social Work Profession*. Retrieved July 2019, from International Federation of Social Workers: https://www.ifsw.org/what-is-social-work/global-definition-of-social-work/

International Finance Corporation, World Bank Group. (2014). *Micro, Small, and Medium Enterprise Finance Improving Access to Finance for Women-owned Businesses in India*. New Delhi: International Finance Corporation.

Iyer, L. (2007). *Women Entrepreneurs -Challenges and Strategies*. New Delhi. : Friedrick E.Bert Stifting.

Iyer, L., Khanna, T., & Varshney, A. (2011, October 18). Caste and Entrepreneurship in India. *Working Paper 12-028*. Harvard Business School.

Jackson, S. L. (2009). *Research Methods and Statistics: A Critical Thinking Approach*. Belmont: Wadsworth, Cengage Learning.

Jain, A. (2016). Green Human Resource Management Concept In Indian Industries – Present Scenario. *International Journal of Multidisciplinary Allied Research Review and Practices*, 171-181. Retrieved 2017

Jakobsh, D. R. (2004). Barriers to Women's Leadership. (T. Oaks, Ed.) *Encyclopedia of Leadership*, 77-81.

Jifiti, T. (2016, April 20). *Catching Up with the Elderly: Optimizing eCommerce for the Golden Agers*. Retrieved from Jifiti. Be Gifted.: https://gifts.jifiti.com/blog/gifting-data/optimizing-for-the-elderly/

Johnson, A. (2000). *The Blackwell Dictionary of Sociology* (2nd ed.). Oxford: Blackwell.

Juneja, P. (2018). *Human Resource Management (HRM) - Definition and Concept*. Retrieved August 2018, from Management Study Guide: https://www.managementstudyguide.com/processes-in-human-resource-management.htm

Jurkowski, E. T., & Keefe, R. (Eds.). (2013). *Handbook of Public Health Social Work*. New York: Springer Publishing Company.

Kait, K. (2013, December 18). *Articles: Legal Services India.* Retrieved from Legal Services India: http://www.legalservicesindia.com/article/article/dr-b-r-ambedkar%E2%80%99s-role-in-women-empowerment-1611-1.html

Kannan, S. (n.d.). *Vedas and Modern Business Management Practices* . Retrieved from Ved Vikas: http://vedvikas.com/samplechapter.pdf

Kapila, U. (2007). *Indian Economy: Performance and Policies.* New Delhi: Academic Foundation.

Kaplan, M. (2013, December 16). *Ecommerce Merchants Should Not Ignore Older Shoppers.* Retrieved from PracticalEcommerce: http://www.practicalecommerce.com/articles/61984-Ecommerce-Merchants-Should-Not-Ignore-Older-Shoppers

Kaul, R. (2018, June 07). From 167 to 130, India's maternal mortality rate down by 22%. *Hindustan Times.*

Kaushik, D. S. (2009). *Women Entrepreneurs.* Jaipur: Ritu Publications.

Keer, D. (1954). *Dr. Ambedkar: Life and Mission* . Bombay: Popular Prakashan.

Kellner, D. (2002). Theorizing Globalization. *Sociological Theory, 20*(3), 285-305.

Khanka, S. S. (2006). *Entrepreneurial Development.* New Delhi: S.Chand Publications.

Kidwai, N. L. (2015). *30 Women in Power: Their Voices, Their Stories.* Faridabad: Thomson Press India Ltd.

Kirkwood, J., & Tootell, B. (2008). Is entrepreneurship the answer to achieving work–family balance? *Journal of Management & Organization*, 285 - 302 .

Kirschenhofer, F. (2006, January 13-14th). Long-term performance of habitual entrepreneurs – which direction to go? *Paper presented at the Workshop on Firm Exit and Serial Entrepreneurship at the Max Planck Institute.* Jena, Germany.

Knopik, M. S., & Moerer, T. (2008). Women Leaders and Entrepreneurs: Learning from One to Teach the Other. *Journal of Women in Educational Leadership*, 137-153.

Knopik, M. S., & Moerer, T. (2008). Women Leaders and Entrepreneurs: Learning from One to Teach the Other . *Journal of Women in Educational Leadership*, 137-153.

Kothari, C. R. (2004). *Research Methodology: Methods and Techniques* (Second ed.). New Delhi: New Age International.

Kraisonswadi, N. (1989). *Women Executives: A Sociological Study in ROle Effectiveness.* Jaipur: Rawat Publications.

Krishnan, J. (2013). *Breaking Barriers.* Mumbai: Jaico Publishing House.

Kumar, A. (2012). *Entrepreneurship: Creating and Leading an Entrepreneurial Organization.* Noida: Dorling Kindersley (India) Pvt. Ltd.

Kumar, B. (Ed.). (2009). *Women Empowerment and Sustainable Development.* New Delhi: Regal Publications.

Kumar, R. (2007). *Ambedkar and His Writings: A Look for the New Generation.* New Delhi: Gyan Publishing House.

Kumari, S. (2012). Women Entrepreneurship in India. *International Indexed & Referred Research Journal*, 6.

Lal, A. T. (2014, December). *Women Entrepreneurs in India - Over the Years!* Retrieved from SSRN: https://ssrn.com/abstract=2176377

LalithaDevi, U. (1982). *Status and Employment of Women in India.* New Delhi: B.R. Publishing Corporation.

Lavanya, T. (2010). *Women Empowerment Through Entrepreneurship.* New Delhi: New Century Publications.

Laxmi, C. S. (Ed.). (1998). *Development of Women Entrepreneurship in India.* New Delhi: Discover Publishing House.

Lewis, K. V., Henry, C., Gatewood, E., & Watson, J. (Eds.). (2014). *Women's Enterpreneurship in the 21st Century: An International Multi-level Research Analysis.* Massachusetts: Edward Elgar Publishing Inc.

Limaye, C. (1999). *Women: Power and Progress.* New Delhi: B. R. Publishing Corporation.

Littunen, H. (2000). Entrepreneurship and the characteristics of the entrepreneurial personality. *International Journal of Entrepreneurial Behaviour & Research*, 295–309.

Lokhande, D. M. (2007). A Study of Women Entrepreneurship Development in Industries Backward Areas. *Entrepreneurship for Economic Development Conference* (pp. 1-8). Jalna: Jain Printers.

Loney, M. (1983). *COmmunity Against Government: The British Community Development Project 1968-1978 A Study of Government Incompetence.* London: Heinemann.

Long, D. D., Tice, C. J., & Morrison, J. D. (2012). *Macro Social Work Practice: A Strengths Perspective* . Boston: Cengage Learning.

Lussier, R. N., Corman, J., & Kimball, D. (2014). *Entrepreneurial New Venture Skills.* New York: Routledge.

Malhotra, S. (2015, June 07). *Invisible Hands.* Retrieved from Business Today: http://www.businesstoday.in/magazine/features/child-labour-in-india-how-it-being-hidden-from-authorities/story/219448.html

Manfredi, S., & Holliday, M. (2004). *Work-Life Balance: An Audit of Staff Experience at Oxford Brookes University* . Oxon, Great Britain: The Centre for Diversity Policy Research.

Mathew, R. V., & Panchanatham, N. (2011, July). AN EXPLORATORY STUDY ON THE WORK-LIFE BALANCE OF WOMEN ENTREPRENEURS IN SOUTH INDIA. *Asian Academy of Management Journal, 16*(2), 77–105.

Maxwell, A. G., & McDougall, M. (2004). Work-life Balance: Exploring the connection between levels of influences in UK Public Sector. *Public Management Review*, 377-393.

Mays, G. P., & Novick, L. F. (Eds.). (2005). *Public Health Administration: Principles for Population-based Management.* Burlington, Massachusetts: Jones & Bartlett Learning.

McClelland, D. C. (1961). *The Achieving Society.* New York: Free Press.

McClelland, D. C. (1967). Money as motivator: some research insights. *McKinsey Quarterly* , pp. 10–21.

McClelland, D. C., Clark, R. A., Roby, T. B., & Atkinson, J. W. (1958). The Effect of the Need for Achievement on Thematic Apperception. In J. Atkinson, *Motives in Fantasy, Action and Society* (pp. 64-82). Princeton, NJ: Van Norstrand.

McCormack, J. (2007). Globalization and Social Work Education and Practice: Exploring Australian Practitioners' Views. *The Journal of Sociology & Social Welfare*, 123-142.

McKay, R. (2001). Women Entrepreneurs: Moving Beyond Family and Flexibility. *International Journal of Entrepreneurial Behavior & Research*, 148 - 165. Retrieved February 2018

McKenzie, W. (2005). *Multiple Intelligences and Instructional Technology.* Virginia: International Society for Technology in Education.

McKenzie, W. (2005). *Multiple Intelligences and Instructional Technology.* International Society for Technology in Education.

Mehta, B., & Parekh, S. (2014). Women Entrepreneur: A Road to Empowerment. *Strategies for Business Challenges*, 71 - 75.

Mehta, B., & Parekh, S. (2014). Women Entrepreneurs: Leading the Way Ahead. *Journal of Social Welfare and Management, 6*, 35 - 40.

Menon, N. (2015, April 10). *Worker (Census Definition).* Retrieved September 21, 2018, from IES Arthapedia: http://www.arthapedia.in/index.php?title=Wroker_(Census_Definition)

Mishra, & Puri. (2006). *Indian Economy* (24th ed.). New Delhi: Himalaya Publishing House.

Mishra, P. D. (1992). *Social Work - Philosophy and Methods.* New Delhi: Inter India Publications.

Mishra, S. (2009, September 11). *Women Enterpreneurship Development in India.* Retrieved from IndianMBA: http://indianmba.com/Faculty_Column/FC1073/fc1073.html

Moerer-Urdahl, T. (2005). A longitudinal qualitative study of collegiate mentoring experiences in the Nebraska Human Resources Research Foundation. Doctoral dissertation, The University of Nebraska-Lincoln.

Moharana, S., & Dash, C. R. (1996). *Entrepreneurship Development.* Jaipur: RBSA Publishers.

Mondal, P. (2016). *Status of Women in Vedic and Post-Vedic Period.* Retrieved from Your Article Library: http://www.yourarticlelibrary.com/society/status-of-women-in-vedic-and-post-vedic-period/4397/

More, V. G. (2011, August). Dr. B. R. Ambedkar"s Contribution for Women's Rights. *Variorum - Multi- Disciplinary e-Research Journal, 02*(I).

Mundra, S., & Sharma, J. (2013). A Comparative Analysis of the Status of Women Entrepreneurship: India V/S Global. *Research Journal of Social Science & Management*, 147-156.

Murthy, B., Chandrashekhar, M., & Rao, M. G. (1986). Entrepreneurial Process and Promises. *Precision.*

Murthy, N. (2003). The Essence of Leadership. *The Smart Manager*. Retrieved February 2017

Nalini, M. (1993). Caste and Patriarchy: Ambedkar's Insight into the Status of Indian Women. In K. S. Chalam, *Relevance of Ambedkarism in India* (pp. 129-138). Jaipur: Rawat Oublications.

Nambiyar, S. (2017). *Success Stories of Entrepreneurs of the Banyan City.* Ahmedabad: University Granth Nirman Board.

Nathani, K. (2019). *#9 Powerful Quotes from Successful Indian Women Entrepreneurs.* Retrieved May 2019, from Entrepreneur India: https://www.entrepreneur.com/article/298133

Nayyar, P., Sharma, A., Kishtawaria, J., Rana, A., & Vyas, N. (2007). Causes and Constraints by Women Entrepreneurs in Entrepreneurial Process. *Journal of Social Science, 14*(2), 99–102.

Newar, S. (2011, January 24). *Mother –The True Identity of Woman.* Retrieved from Agniveer: http://agniveer.com/mother-woman/

Nieuwenhuizen, C. (2009). *Entrepreneurial Skills.* Lansdowne, Cape Town: Juta and Company Ltd.

Omvedt, G. (2008). *Ambedkar: Towards an Enlightened India.* U.K.: Penguin.

(2005). *Opportunities for Women Entrepreneurship.* New Delhi: National Institute for Industrial Research.

Pandya, R. (2016). *Skill Development and Entrepreneurship in India.* New Delhi: New Century Publications.

Pandya, R. (2016). *Skill Development and Entrepreneurship in India.* New Delhi: New Century Publications.

Pareek, U., & Rao, T. V. (1992). *First Handbook of Psychological and Social Instruments.* New Delhi: Concept Publishing Company.

Park, K. (2017). *Parks Text Book Of Preventive & Social Medicine* (24 ed.). Jabalpur: Banarasidas Bhanot.

Patel, C. (2013, April). A Study of Women Entrepreneurs of Vadodara City. Vadodara.

Patel, J. R. (2005, May). *Growth of Women Entrepreneurship in Baroda.* Vadodara, Gujarat, India.

Patscher, K., Wade, L., & Simon, P. (2010). *Engaging Philanthropy to Promote Gender Equality and Women's Empowerment.* New York: United Nations Headquarters. Retrieved March 28, 2017

Pearce, C. L., Maciariello, J. A., & Yamawaki, H. (2010). *The Drucker Difference: What the World's Greatest Management Thinker Means.* Pennsylvania: McGraw Hill.

PHD CHAMBER OF COMMERCE AND INDUSTRY . (2019, March). *Women Entrepreneurship: Transforming from Domestic Household to Financial Independence.* Retrieved from PHD Research Bureau: https://www.phdcci.in/wp-content/uploads/2019/03/Women-Entrepreneurship-Transforming-from-domestic-household-to-financial-independence.pdf

Radhakrishnan, & Moore. (1957). Vedas and Upaniṣads. In Phillips, *Philip's Encyclopaedia* (p. 3). London: Phillips.

Rajagopalan, R. (2016). *Environmental Studies: From Crisis to Cure.* London: Oxford University Press.

Rajarajeswari, L. (2012). *Women Empowerment - The Condition and Position of Women.* Retrieved from IndianMBA: http://www.indianmba.com/Faculty_Column/FC1435/fc1435.html

Ramamohan, C. V., Samineni, S., & Vemuri, S. (2011). *EMPOWERMENT OF WOMEN THROUGH LEADERSHIP & ENTREPRENEURSHIP.* New Delhi: Himalaya Publishing House.

Ramsay, S., & Boddy, J. (2017). Environmental Social Work: A Concept Analysis. *The British Journal of Social Work, 47*(01), 68-86.

Rani, D. L. (1996). *Women Entrepreneurs.* New Delhi: A.P.H Publishing Corporation.

Rao, B. S. (2007). *Women Empowerment & Planning Process.* Retrieved March 31, 2017, from Content Writer India: www.contentwriter.in/articles/others/women-empowerment.htm

Reddy, P. N. (1991). Problems of Women Entrepreneurs in Goa: A Pilot Study. *Khadi Gramadyog, 37*(4), 157–159.

Richardson, D., & Robinson, V. (Eds.). (1993). *Introducing Women's Studies.* London: Macmillan Press.

Rizvi, A. F., & Gupta, K. L. (2009). Women Entrepreneurship in India – Problems and Prospects. *OORJA Journal of Management and I.T., VII*(2), 35–41.

Roy, R. (2011). *Entrepreneurship.* New Delhi: Oxford University Press.

Roy, T. (2001). WOmen Entrepreneurs in Urban Informal Sector. *TISS Mumbai.*

Rusinko, C. A. (2010). Integrating Sustainability in Higher Education: A Generic Matrix. *International Journal of Sustainability in Higher Education*, 250-259.

Saha, J. (2006). *Management and Organizational Behaviour.* New Delhi: Excel Books India.

Sangolagi, K., & Alagawadi, M. (2016). Women Entrepreneurs . *International Journal of Advancement in Engineering Technology, Management and Applied Science*, 216-222.

Sarada, S. (2004). *Stress of Women Entrepreneurs: A Comparative Study with Women in Selected Groups in the District of Ernakulam.* Kochi: School of Management Sudies, Cochin University of Science and Technology.

Saurabh, S. (n.d.). ISSUES AND CHALLENGES FACED BY WOMEN ENTREPRENEURS AND THEIR TRAINING NEEDS. *International Journal of in Multidisciplinary and Academic Research (SSIJMAR)July-August (ISSN 2278 – 5973), 1*(2).

Schumpeter, J. A. (1955). Economic Theory and Entrepreneurial History. In H. Aitken, *Explorations in Enterprise* (pp. 51-52). Cambridge: Harvard University Press.

Schwartz, E. B. (1973). Can Women Compete in Small Business? *Journal of Small Business Management* , 4.

Sehgal, R. (2005). Social Work and Industry- Dilemma of Partnership. *Contemporary Social Work*, 66-95.

Sellitz, C., Jahoda, M., Deutsch, M., & Cook, S. W. (1951). *Research Methods in Social Relations.* New Tork: Holt, Rinehart and Winston.

Setty, E. D. (1980). Developing Enterpreneurship Among WOmen. *Man and Development, II*(03).

Setty, E. D., & Krishnamoorty, P. (1981). *Strategies for Developing Women Entrepreneurship.* New Delhi: Akanksha Publishing House.

Shah, H. (2013). *Creating and Enabling Environment for WOmen Entrepreneurship in India.* New Delhi: Economic and Social Commission for Asia and the Pacific (ESCAP). Retrieved October 2019, from https://www.unescap.org/sites/default/files/ESCAP-SSWA-Development-Paper_1304_1.pdf

Shane, S., & Venkataraman, S. (2000). The Promise of Entrepreneurship as a Field of Research. *Academy of Management Review*, 217-226.

Sharma, D. D., & Dhameja, S. K. (2010). *Women Entrepreneurs and Enterprise Management.* Chandigarh: Abhishek Publications.

Sharma, J. P. (2009). *Environmental Studies.* New Delhi: Laxmi Publications.

Sharma, K. (2016, May). *14 Percent of Indian Businesses Run By Women Entrepreneurs.* Retrieved October 2017, from Indian Web2: Technology, Web and Start-ups: https://www.indianweb2.com/2016/05/10/14-percent-indian-businesses-run-women-entrepreneurs/

Sharma, M. (2017). Women Entrepreneurs: A Study of Acheivement Motvation, Leadership and Work-life Balance of Women Entrepreneurs of Gujarat. *Concept Paper*. Unpublished.

Sharma, P. (2014). Changing Status of Women-owned Enterprises in IndiaAn. *Global Journal of Finance and Management*, pp. 933-944.

Sharma, S. (2016). *Entrepreneurship Development.* New Delhi: PHI Learning Pvt. Ltd.

Shekhar, N., & Padmadeo, K. B. (2009). *Women Entrepreneurship and Empowerment.* (B. Kumar, Ed.) New Delhi: Regal Publications.

Shelton, L. M. (2006). Female Entrepreneurs, Work–Family Conflict, and Venture Performance: New Insights into the Work–Family Interface. *Journal of Small Business Management*, 285-297. Retrieved February 2018

Sheopuri, A., & Sheopuri, A. (2015). Green Hr Practices in the Changing Workplace. *Business Dimensions*, 13-26.

Siddiqui, S. (2008). *Women Enterpreneurs in Export Trade.* New Delhi: Regal Publications.

Singariya, M. R. (2014). Dr. B. R. Ambedkar and Women Empowerment in India. *Journal of Research in Humanities and Social Science*, 01-04.

Singh, A. (2014). Role of Women Entrepreneurs in India: A SWOT Analysis. *International Journal of Management and International Business Studies*, 231-238.

Soota, A., & Gopalan, S. R. (2016). *Entrepreneurship Simplified: From Idea to IPO.* UK: Penguin.

Soundarapandian, M. (1999). *Women Enterpreneurship: Issues and Strategies.* New Delhi: Kanishka Publishers and Distributors.

Stevenson, M. (2018, March). Employee Engagement to Employee Experience: Why HR Should focus on the Individual? *HR Exchange Network*. Retrieved August 2018

Swarnalatha, C., & Rajalakshmi, S. (2016). *A HAND BOOK ON WORK-LIFE BALANCE.* Solapur: Laxmi Book Publication.

Tang, S. (2010, June). *Work support, work-family enrichment, work demand and work well-being among Chinese employees : A study of mediating and moderating processes.* Retrieved from Digital Commons @ Lingnan University: http://commons.ln.edu.hk/do/search/?q=author_lname%3A%22TANG%22%20author_fname%3A%22Shuwen%22&start=0&context=2433414

Taylor, S. H., & Roberts, R. W. (2013). *Theory And Practice of Community Social Work* . Jaipur: Rawat Publication.

Teasley, D. (n.d.). *Achievement Motivation: Theory & Definition.* Retrieved from Study.com: http://study.com/academy/lesson/achievement-motivation-theory-definition-quiz.html

Thomas, R. M. (2003). *Blending Research Methods in Theses and Disstertations.* California: Sage Publications Inc.

Tiwari, A. (2007). *Women Entrepreneurship and Economic Development.* New Delhi: Sarup & Sons.

Tiwari, A. K. (2016, April 26). Walmart lends a helping hand to women entrepreneurs. *DNA*.

Tiwari, S., & Tiwari, A. (2007). *Entrepreneurship Development in India.* New Delhi: Sarup and Sons.

TNN. (2016, May 03). Women Entrepreneurs Come Together for Better Business. *The Times of India*, p. 1.

Vaidya, S. (2014). *Developing Entrepreneurial Life Skills: Creating and Strengthening Entrepreneurial Culture in Indian Schools.* New Delhi: Springer Science & Business Media.

Valeri, A., Parton, B., & Robb, A. (2014). *Entrepreneurship Education and Training Programs around the World: Dimensions for Success.* Washington, DC: World Bank Publications.

Vasanthagopal, R., & Santha, S. (2008). *Women Entrepreneurship in India.* New Delhi: New Century Publications.

Vinay, D., & Singh, D. (2015). Status and Scope of Women Entrepreneurship. *Universal Journal of Management* , 43-51.

Virtanen, M. (1997). The role of different theories in explaining entrepreneurship. (S. Kunkel, Ed.) *Entrepreneurship: the engine of global economic development.*

Walker, E., Wang, C., & Redmond, J. (2008). Women and work-life balance: is home-based business ownership the solution? *Equal Opportunities International*, 258 - 275. Retrieved December 2017

Warke, N. P., & Narkhede, P. A. (2014). WOMEN EMPOWERMENT: SKILLS ENHANCEMENT THROUGH ENCOURAGING ENTREPRENEURSHIP. *Tactful Management Research Journal*, 70-75.

Women's Entrepreneurship Development (WED) Programme. (n.d.). Retrieved from International Labour Organization: http://www.ilo.org/empent/areas/womens-entrepreneurship-development-wed/lang--en/index.htm

Women Entrepreneurs: A Study of Achievement Motivation, Leadership and Work-life Balance of Women Entrepreneurs of Gujarat

Introduction

According to P. F. Drucker an entrepreneur is one who always searches for change, responds to it and exploits it as an opportunity."

According to Cantillion "entrepreneur is the agent who buys means of production at certain prices, in order to sell at prices that are certain at the moment at which he/she commits himself/ herself to his/ her cost".

There exist a number of schools of thought, which view the notion of entrepreneurship from fundamentally different perspectives. The term has been used to define a wide range of activities such as creation, funding, adapting, and managing and venture.

Recently a new trend has emerged in India wherein women are venturing as entrepreneurs and are contributing to the economic development of the country. Women are increasingly seeking entrepreneurship as an avenue for economic growth. The government is also playing a significant role in mobilising women to become entrepreneurs through its different programs. This is helping is the number of women entrepreneurs increase much more.

Women Entrepreneurs

The concept of women entrepreneurship is becoming a global phenomenon. Women access to knowledge, skills, resources, opportunities and power but still remain rather low especially in the rural areas. Although women remain involved in family trade but they are neither in organized form nor do women have its ownership making their role as a mere helper. Women of today are clearer in their life-goals. They know how to undergo all the hardships as well as how to cherish the success which comes after. In our culture, women are socially conditioned to be submissive and not speak much. But a woman with entrepreneurial aspirations must make herself heard. While making tall claims on achieving women empowerment, we must also provide opportunities and the right kind of platforms to women to achieve.

Women have been taking increasing interest in recent years in income generating activities, self employment and entrepreneurship. This is seen in respect of all kinds of women both in urban and rural areas. Women are taking up both traditional activities and also non-traditional activities. It is clear that more and more women are coming forward to set up enterprises. In the process of entrepreneurship, women have to face various problems associated with entrepreneurship and these problems get doubled because of her dual role as a wage earner and a homemaker. It becomes all the more important to study what keeps these women well motivated in order to achieve their set goals and targets. It is also important to study what leadership patterns are followed by these women in order to handle their human resources and other resources as well as how they balance their work-life while setting up their enterprises.

Social Work Significance

The profession of social work has always emphasized on empowering women and it is considered to be one of the most important thrust areas of the profession. Encouraging livelihoods and promoting skill-development has always been of prime importance to the field of social work. Every trained social worker aims to uplift the status of their stakeholders in every possible way. Promoting gender equality and encouraging gender-neutral workplaces and work environment is the primary principle of social work practice. With the dawn of the new century, women have sought to look more beyond just a stable job. The concept of women entrepreneurship is becoming a global phenomenon. Women access to knowledge, skills, resources, opportunities and power but still remain rather low especially in the rural areas. Although women remain involved in family trade but they are neither in organized form nor do women have its ownership making their role as a mere helper. Women of today are clearer in their life-goals. They know how to undergo all the hardships as well as how to cherish the success which comes after. In our culture, women are socially conditioned to be submissive and not speak much. But a woman with entrepreneurial aspirations must make herself heard. While making tall claims on achieving women empowerment, we must also provide opportunities and the right kind of platforms to women to achieve.

An entrepreneur plays a very dynamic role for his enterprise. A woman, considered vulnerable, has to combat a lot of hurdles and obstacles which come in her way achieving her goals. The researcher will also consider the environmental challenges, strengths and weaknesses of women entrepreneurs and that how under even stressful times, a woman exhibits strength in holding her career and personal life together.

Duality of the roles of Women

Women entrepreneurs have to play a dual role in their lives. They are supposed to undertake household chores as well as manage the enterprise. If a woman seems to be too ambitious towards her career, she is labelled as insensitive towards her homely responsibilities. Women forget that they too have a capacity in trying to fulfil their dual roles. This feeling of being a super-woman, gives way to conflicts. Career women often have to face work-home role conflict and so the concept of work-life balance becomes an important parameter to be studied while researching on women entrepreneurs.

Working women and especially enterprising women need to be authoritative and showing some leadership skills. It also becomes necessary to study the leadership aspect of enterprising women as that factor helps them retain manpower and expand their enterprise. The researcher will also study the leadership styles adopted by enterprising women and how it affects their entrepreneurial performance.

Women entrepreneurs represent a group of women who have broken away from the beaten track and are exploring newer avenues of economic participation. Women with high drive, creativity, innovation and those who are keen on achieving on their own have taken up the challenging role of entrepreneurship. Women no longer confine themselves to conventional fields like embroidery, knitting and tailoring but are now venturing into new fields like electronics, pharmaceuticals engineering and services. These women are mostly urban, appropriately educated, sometimes having specialised skills and equipping themselves with some training and financial support. In spite of women taking entrepreneurship in many challenging fields the current entrepreneurial activity in India is not very high. There exist a definite gap between the current status of women entrepreneurial activity and their potential. The social origins of women entrepreneurs play an important role in defining the kind of work they undertake. Women also find micro enterprises attractive because of low barriers to entry and the flexible nature of work which makes it easy to combine gainful employment with domestic responsibilities. (Carr, 1990)

Women Entrepreneurs in Gujarat

Business runs in the blood of every Gujarati. It has spread its acumen over USA, Canada and many other Western horizons. The Centre for Entrepreneurship Development (CED) was established in 1970 in Gujarat to enhance trader's expertise and help the businesses flourish. In order to encourage more women entrepreneurs from the state, CED started various programmes to train women in this domain.

The Government of Gujarat has taken many steps to encourage the women of the state to take up their entrepreneurial careers. While he was the Chief Minister of the state, Shri Narendra Modi kept on emphasizing in his speeches that entrepreneurship is inbuilt in the people of Gujarat. He further kept addressing that women play an important role for the economic development. According to him, even in the dairy sphere, the contribution of men is minimal. He suggested that women should be made an integral part in the decision making process in almost all the spheres.

High need for achievement, high need for autonomy, high need for creativity, and prevalence of initiative, industriousness, and foresight through self-reliance and possession of managerial skills are all effective pointers to an entrepreneur. (Sarada, 2004). In this globalised and digitalised era, many women have found their place in the market as entrepreneurs. With the changes in the way a business runs, it becomes important to study how motivated women are to pursue their dream of enterprising. There is a great scope to study the work-life balance and leadership skills of the women of the 21st century as this era is an era of continual transformation and change and entrepreneurs have to work up to their mark and even beyond to survive in an otherwise competitive market. The Government also has rolled out the "Start-up India" initiative to encourage more and more entrepreneurs in the various business sectors.

Social Work is an inclusive profession. Its core values speak of inclusive development and acceptance of people. Women have always borne the brunt of social changes and therefore labelled as vulnerable. However, in this fast-paced world, the challenges are of a different kind. It has been a long pending economic independence that women have achieved on their own merit. Former leaders of the country like Mahatma Gandhi and Dr. B. R. Ambedkar had always encouraged participation of women in the freedom struggle and economic development of the country. Despite this ideology, it took years together for women to achieve this freedom. In recent years, women have started being more and more assertive about their economic independence and many such women are turning to entrepreneurship. They are now their own bosses. Women may have reached to the pinnacle of success but not all success comes easy. Due to the dual social roles that women play, success would not have been possible without sacrificing their family life or other social expectations. It is with this intention that this study needs to take place so that the expectations, ideas and aspirations of the enterprising women of the millennium can be understood and appropriate assistance can be provided to foster their skill, capacity and talent.

Objectives of the study

1. To study the socio-economic background of women entrepreneurs in Gujarat
2. To analyse motivational factors behind women entrepreneurship
3. To analyse the leadership quality among women entrepreneurs in Gujarat.
4. To analyse the degree of work-life balance that women entrepreneurs face and its effect on their entrepreneurial performance.
5. To suggest a framework to promote women entrepreneurship in Gujarat.

Operational Definition:

Study title - Women Entrepreneurs - A Study of Achievement Motivation, Leadership and Work-life Balance of Women Entrepreneurs of Gujarat.

Women Entrepreneurs - Women entrepreneurs represent a group of women who have broken away from the beaten track and are exploring newer avenues of economic participation. (Rani, 1996)

Achievement Motivation - The concept of achievement motivation has been used to describe individual efforts towards personal goals in social surroundings (Cassidy & Lynn, 1989). It is also found to be a typical characteristic among individuals who are successful in their work (McClelland, The Achieving Society, 1961) (McClelland, Money as motivator: some research insights., 1967). A strong achievement motivation is probably the one characteristic most clearly describing the attitudes and the behaviour of an entrepreneur (Virtanen, 1997).

Leadership - Leadership is a process whereby an individual influences a group of individuals to achieve a common goal. Leadership style is characterized by the consistent patterns of exhibited behaviors, attitudes, and values. Leadership has been defined as traits, behaviors, influence, persuasion, interaction patterns, role relationships, and as administrative position; and most definitions assume that leadership involves a process whereby intentional influence is exerted by one to others in an attempt to guide relationships (Moerer-Urdahl, 2005).

Gender is integral to contemporary notions of effective styles that have morphed from a traditional masculine, autocratic style to a more feminine and androgynous style of democratic and transformational leadership (Knopik & Moerer, 2008).

Work-life Balance - Work-Life Balance does not mean an equal balance. It means the capacity to schedule the hours of professional and personal life so as to lead a healthy and peaceful life. It is not a new concept. It emphasizes the values, attitudes and beliefs of women regarding their age to work in organizing and balancing their work and personal life. (Deshmukh & Deshmukh, 2016)

Women turn to entrepreneurship for various reasons especially for satisfying their personal needs of self-expression an autonomy. This also makes their lives increasingly complex with pressure, intense working hours and lesser involvements in the family; hence they juggle to satisfy both the roles. (Rani, 1996) They may receive additional help at home in the form of servants, parents, in-laws and husband but still the entire responsibility of the home rests with the woman. But the problems remain, as the entrepreneur is not a super woman and there follows a conflict.

Relevance of this study to Social Work Profession

Social Work is an inclusive profession. Its core values hold inclusive development and acceptance of people. Women have always borne the brunt of social changes and therefore labelled as vulnerable. However, in this fast-paced world, the challenges are of a different kind. Women may have reached to the pinnacle of success but not without sacrificing their family life or other core values. It is with this intention that this study needs to take place so that the expectations, ideas and aspirations of the enterprising women of the millennium can be understood and appropriate assistance can be provided to foster their skill, capacity and talent.

Research Methodology:

Who is a woman entrepreneur? This question has been defined in different ways by different people at different points of time. The Government of India considered the majority participation of women in enterprises as a necessary qualification for calling such enterprises as women-enterprises. Recently, the term majority participation has been modified to majority equity participation of women in organizations. This again, could be debated as many times, women may invest but not as actively participate in the running of the enterprise. It has been thus thought that both equity participation as well as participation in the running of enterprises to a greater extent by women entrepreneurs should be a necessary criteria for calling enterprises as women-run enterprises.

Therefore, for this study, the women entrepreneurs considered were based on the following **inclusion criteria**:

1. The woman entrepreneur should have contributed to the equity capital of the concern.

2. The woman entrepreneur should be actively participating in the running of the organization and in all functional areas of management.

3. The establishment of the enterprise be minimum three years.

4. The establishment must help generate income for all those employees associated with it.

Discussions with appropriate authorities at the government departments like District Industries Centres (DICs) and Centres for Entrepreneurship Development (CEDs) as well as the Chambers of Commerce and Industries and Entrepreneurship Development Institute (EDI) were done. With the preliminary information provided, an attempt to formulate a list of women-run enterprises was created. It was felt that as no census data is available on the number of women entrepreneurs in the state, **a snowball sampling method** was followed.

This study follows **an exploratory cum descriptive research design,** as it will help the researcher to understand the issue more thoroughly, before attempting to quantify mass responses into statistically inferable data. Used properly, exploratory research will provide rich quality information that will help identify the main issues that should be addressed in our survey and significantly reduce the level of prejudice. Within the exploratory design, the researcher will incorporate the use of Case studies. One case study from each zone was presented. Reliability on Secondary Sources of Data was also adopted.

Tools of Data Collection:

A properly structured tool for collecting demographic data was constructed using more of close-ended questions as the study aims at obtaining the opinions of its respondents. This tool was developed by referring the various standardised tools available on the relevant subject. A pilot sample was studied and the sample data was treated on Cronbach's Alpha Test to assure of the reliability and validity of the tool.

Universe:

For the purpose of this study, the researcher studied all the women-owned, women-run enterprises within the State of Gujarat. However, four major zones i.e. Ahmedabad, Rajkot, Surat and Vadodara were covered.

Sampling:

Snowball Sampling method was used. In this method of sampling, the existing subjects under study refer the future subjects from among their acquaintances. The research scholar tried to obtain the names of Women Entrepreneurs from the local Chamber of Commerce and Industries from each of the zones under study and proper representation from each zone was selected for the study. All women-owned, women-run enterprises established since the last three years in the cities of Ahmedabad, Rajkot, Surat and Vadodara were studied.

Variables:

Dependent Variables: Work-Life Balance, Motivation and Leadership

Independent Variables: Age, Education, Experience and Income

Hypotheses: 1. Higher the income, higher is the motivation to work.
2. With increased experience in the field, entrepreneurs develop their own leadership.

3. Work-life balance is absent in women entrepreneurs.

Statistical Treatment:

The filled questionnaires were numbered serially and checked for consistency of the data, and any in-consistent data. The questions and responses were codified to represent the data in tabular / graphical form wherever appropriate. The collected data was correlated with the various dependent and independent variables. Hypotheses were verified using appropriate statistical tools and with the help of Excel / SPSS (Statistical Package for Social Science for Windows).

Limitations:

According to the researcher, no research can be a perfect research. There is always a scope of error. In this research, the researcher feels that there may be certain limitations that might affect the study. These limitations are,

- Lack of willingness of the respondent to participate in the research.
- Non-availability of Respondents.
- Lack of understanding about the topic in concern by the respondents.
- Authenticity of the secondary sources of data.

Data Analyses

The structured tool was first treated under the Cronbach's Alpha Test to find out its reliability and validity. On the exposed sample of 30 tools, the following score was found:

Case Processing Summary			
		N	%
Cases	Valid	30	100.0

	Excluded[a]	0	.0
	Total	30	100.0

a. List-wise deletion based on all variables in the procedure.

Reliability Statistics

Cronbach's Alpha	Cronbach's Alpha Based on Standardized Items	No of Items
.849	.850	55

As per the obtained score, it can be said that the designed tool was highly reliable for its use and it would measure what it intends to measure.

From the analysed data on 15 tools, the following has been inferred:

Socio-economic Profile

- 59 (37.3%) respondents fall between the age groups of 31-40 years.
- 71 (44.9%) respondents have pursued education up to Graduation level whereas 52 (32.9%) of the respondents have pursued education up to Post Graduate level.
- Majority of the respondents i.e. 77.2% are married.
- 54 (34.2%) respondents have an experience of more than 10 years in running an enterprise.
- 60 respondents, i.e., 38% have an experience of 2-5 years in running their current enterprise whereas 38 (24.1%) respondents have an experience of running their own enterprise for the last 6-10 years.
- 41 respondents, i.e., 25.9% have an average monthly income of Rs. 21,001 - 50,000. A sizeable number of respondents, 32 (20.3%) have answered that their monthly income is not regular.

- 43 respondents' husbands have their own business i.e. 27.2% and an equal per cent belong to professional category. Professions like Chartered Accountants, Doctors etc. are considered a part of the Professional category.
- While 39 respondents, i.e., 25.3% have chosen not to disclose their husband's average monthly income, 24.7% respondents have claimed that the average monthly income of their husbands is more than Rs. 50,000/-.
- 57 respondents, i.e., 36% have husbands whose minimum qualification is graduation.
- Majority of the respondents, i.e., 81.6% have 3-5 members in their families.
- 51 respondents, i.e., 32.3% have only one child.

On the scale of **Achievement Motivation**, the following inferences were made -

Table 2.1.1 Cumulative response on Achievement Motivation

N=158

Achievement Motivation		
	Frequency	**Percent**
Mostly disagree	15	9.5%
Agree to some extent	90	57%
Mostly agree	53	33.5%
Total	**158**	**100%**

The above graph shows the cumulative responses of the respondents on Achievement Motivation. It shows that out if the 158 respondents, majority of the respondents, i.e., 57% agree to some extent on having Achievement Motivation, 33.5% mostly agree that they have high Achievement Motivation and 9.5% of respondents disagree on having Achievement Motivation.

Graph 2.1.1 Showing Achievement Motivation of Respondents

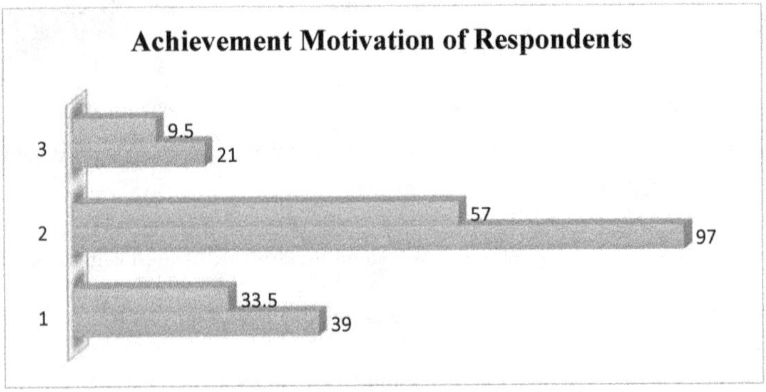

Thus, it can be concluded that there is visible level of achievement motivation present among the respondents. It is this motivation that aspires women to work harder and achieve greater heights in their career. There is a great sense of achievement in women, and on knowing that people like them have achieved much in life, they are also motivated to work harder.

This also hints at the aspiring and hardworking nature of the women respondents.

Respondents own their actions and believe that they themselves like to deal with difficult situations so that the blame or praise comes to them alone.

On the scale of **Leadership**, the following trends were observed -

Table 3.1.1 Showing Cumulative responses on Leadership
N=158

	Frequency	Percent
Sometimes	39	24.7
Always	119	75.3
Total	**158**	**100.0**

Since none of the responses hint towards negation in Leadership, the 'Never' option has been deleted for easy understanding, The above table can be graphically represented as:

Graph 3.1.1 Showing Cumulative responses on Leadership.

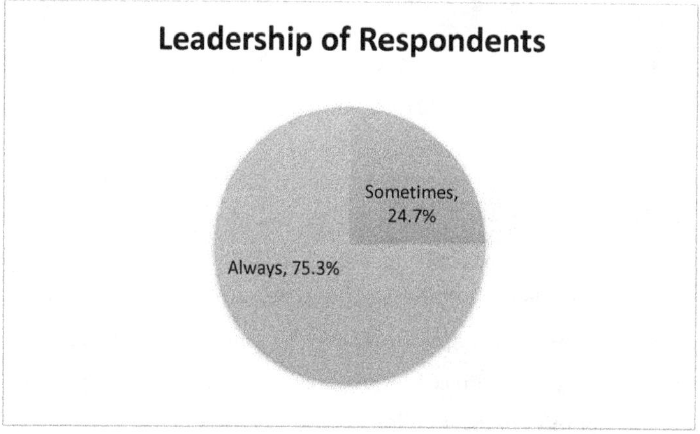

From the data, it can be said that most of the women accept that they closely monitor the tasks at hand.

Many respondents believe in understanding and respecting other peoples' boundaries and limitations thereby knowing how to delegate tasks to them. This also makes them efficient leaders as they understand and accept their team members as they are. A more androgynous cum transformational leadership style is visible among women entrepreneurs.

Respondents feel they need to be more connected with their staff and hence may overlook mistakes however, if the mistakes can lead to serious consequences, they do not hesitate in putting their personal relations apart from their professional ones.

On the scale of **Work-life Balance**, the following inferences were made -

Table 4.1.1 Showing cumulative responses of Work life Balance of Respondents

N=158

	Frequency	**Percent**
If it little true	01	0.6
If it is somewhat true	39	24.7
If it is fairly true	97	61.4
If it is definitely true	21	13.3
Total	**158**	**100.0**

The above table shows that out of the total respondents, i.e. 158, for a majority of respondents, 61.4% achieving work life balance is fairly true, for 24.7% respondents, achieving work life balance is somewhat true, for 13.3% respondents, achieving work life balance is definitely true whereas for only a small population of 0.6%, it is difficult to achieve work life balance.

This can be more descriptively represented through a graph, as below:

Graph 4.1.1 Showing cumulative responses of Work life Balance of Respondents

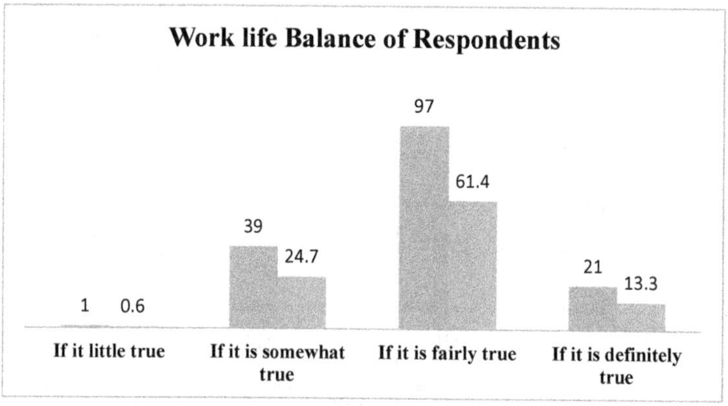

As far as Work-life Balance is concerned, the respondents felt that their personal lives were not affected by their work and they could easily balance it out.

Respondents expressed that they felt a deep sense of satisfaction by contributing to their family and the society at large by being in business and this feeling motivated them to push harder to achieve their goals.

On the basis of the data collected and statistically treated statistical, a collective result on the overall association between Achievement Motivation, Leadership and Work life Balance of Respondents was prepared.

A tabular representation of the same is as below:

Table 5.13 Showing Association between Independent and Dependent Variables.

N=158

		Achievement Motivation	Leadership	Work Life Balance	Age of the Respondent	Education	Experience	Average Income (Per Month)
Achievement Motivation	Pearson Correlation	01	.456**	.367**	.095	.055	.001	.003
	Sig. (2-tailed)		**.000**	**.000**	.235	.493	.994	.974
	N	158	158	158	158	158	158	126

Leadership	Pearson Correlation	.456**	01	.422**	.011	.077	.062	.301**
	Sig. (2-tailed)	**.000**		**.000**	.890	.334	.441	**.001**
	N	158	158	158	158	158	158	126
Work Life Balance	Pearson Correlation	.367**	.422**	01	.197*	.028	.168*	.376**
	Sig. (2-tailed)	**.000**	**.000**		**.013**	.723	**.035**	**.000**
	N	158	158	158	158	158	158	126

(The mean difference is significant at the 0.05 level.)

The above table shows association between the independent and dependent variables. It can be seen that Achievement Motivation is highly significant and associated with Leadership and Work life Balance whereas, it is not significant with any of the Independent Variables.

Leadership is highly significant and associated with the dependent variable of Work life balance and the Independent variable of Average Income (Per Month). It can be interpreted that respondents who have higher income have better leadership skills. However, there is no significance between leadership and experience. Hence, the hypothesis that with greater experience, the respondents develop better leadership skills is rejected. This gives scope to study the leadership patterns of respondents and what factors foster leadership among women entrepreneurs.

Work life Balance is highly significant with Age of the respondents, Experience of running own enterprise and Average Income (per month). This finding emphasizes on the fact that with increased age, respondents try to find their own ways to achieve work life balance. Hence the hypothesis, Work life Balance is absent in Women Entrepreneurs (H_3) cannot be out rightly rejected as during their initial day of start-up, respondents do find it difficult to achieve work life balance, hence it is absent, however, gradually, respondents learn to strike a balance between their work life and their family life.

Work life Balance is also greatly significant with Experience of running own enterprise. With increased experience of running own enterprises, respondents find out ways how they can balance their work life and their personal life.

Work life balance is also highly significant with Average Monthly Income. This shows that as respondents start earning more, they are able to exhibit better leadership in terms of delegating and managing their work in such a manner that they can achieve better work life balance.

The data gives a very vivid picture of how the dependent variables are associated with the independent variables in the current study. Clarity in terms of the interrelatedness of variables is also got from the data set.

Further, the data was also exposed to *f*-test for analysis of variance and the results on the *f*-test were further exposed to post-hoc test to determine the comparison within variables. This led to acceptance of two hypotheses (H_1 & H_2) while non-acceptance of H_3

It has been proven from the obtained data of the present study that income really plays a greater role in motivating the women entrepreneurs into working harder towards their business thereby helping them succeed. The result from the Chi-Square Value for association is 20.699^a and the Significant p-Value obtained is 0.023. **Hence H_1 is accepted.**

It has been proven from the data that leadership is highly significant with experience of the respondents of running their own enterprises within the purview of this study. The correlation between Experience, an Independent Variable and Leadership, a Dependent Variable reveals the Chi-Square Value as 8.627^a and the obtained Significant p-Value as 0.035. Leadership is also dependent upon Income, Occupation of the husband as well as partially with Number of Family Members of the respondents. There is greater significance between Number of Years of Experience of running own enterprise and Achievement Motivation, Leadership and Work life Balance (p-values=0.001, 0.002 and 0.014 respectively)

By applying Post-Hoc Studies using the Scheffe's method it is found that Achievement Motivation is highly significant with respondents having 6-10 years of experience with respondents having 2-5 years of experience (p-value=0.029) as well as those having more than 10 years of experience (p-value=0.002). **Therefore it can be established that H_2 is accepted.**

From the current study, it can be inferred that the women entrepreneurs have a balanced work life and family life. On correlating Work life Balance with the Independent variables, it was found that there is significant association between Age, Experience and Income and Work life

Balance (p-values = 0.013, 0.035 and 0.000 respectively). Even on f-Test, Work life balance is significant with experience (p-value = 0.014) thereby signifying that with more number of years in their own business, women adapt to achieving more Work life Balance; it is also significantly correlated with income, whether it is their own or that of their husband and/ or of the family (p-values = 0.000, 0.008 and 0.001 respectively). From these significant findings it can be concluded that women entrepreneurs do strike a balance between their work life and their family life and that domestic responsibilities do not hamper their entrepreneurial performance and vice versa. **Therefore, H_3 is not accepted.**

A strong conclusion which can be drawn from the quantitative data is that each entrepreneur has their own set of skills and struggles and it takes a great deal of courage to continue working despite any fruitful result visible. A lot of decisions taken while enterprising may not be immediately rewarding but they gradually enhance the capacity of the women to appreciate their own working style and their own actions.

Conclusion

From the above data, it can be concluded that Women Entrepreneurs really have made a big contribution in the nation's economy. They have taken a road less travelled to be able to achieve their goals and aspirations. They have worked tirelessly and relentlessly to achieve their targets despite the society thinking of them as a weaker sex. For them, their gender has never been a matter of shame and their aspirations have not been put down because of their gender. On the contrary, being females, they have received more recognition for being able to run successful businesses besides managing home and family.

As far as **Achievement Motivation** is concerned, Respondents enjoy reading biographies of successful people in order to derive inspiration from their lives thereby making their own work life better. Respondents plan well in time and productively utilize their free time by efficient planning. There is a great sense of achievement in women, and on knowing that people like them have achieved much in life, they are also motivated to work harder. This also hints at the aspiring and hardworking nature of the women respondents. Majority of the women respondents completely agree that they never leave a task unfinished. They continue persevering hard in order to get their tasks done. Women respondents have this innate urge to do something others have not done before an hence, they always compete with themselves to become better at their business. For them there is no substitute to hard work and they give lesser importance to luck. Moreover, they don't wait for people's approval on matters related to their work; if they feel things will work in their favour, they go ahead with their plans. Respondents agreed to befriending the most intelligent people in order to live up to the standards required in business. Respondents own their actions and believe that they themselves like to deal with difficult situations so that the blame or praise comes to them alone.

On the **Leadership** front, respondents have come across to be encouraging to their team members and have tried to foster an environment where even their subordinates learn and take up initiatives. This data also shows their interest in coaching people to be like them. The respondents are satisfied with their team members doing well and not intimidated by them. Nearly 72% of the women accept that they closely monitor the tasks at hand. A fair amount of women respondents i.e. 63% respondents believe in understanding and respecting other peoples' boundaries and limitations thereby knowing how to delegate tasks to them. This also makes them efficient leaders as they understand and accept their team members as they are. When correcting mistakes, 51% respondents do not worry about jeopardizing relationships sometimes. They feel they need to be more connected with their staff and hence may overlook mistakes however, if the mistakes can lead to serious consequences, they do put their personal relations apart from their professional ones. Since most of the respondents are self-made entrepreneurs, they have devised their own ways of leading their team. However, 48% of respondents enjoy reading articles, books and trade journals about their profession, and then implementing the new procedures which they have learned by reading.

As far as **Work-life Balance** is concerned, the respondents felt that their personal lives were not affected by their work and they could easily balance it out. Nonetheless, only 35% respondents are able to meet prescribed deadlines and schedules, without affecting their home life. 35% respondents work for extra hours to get their work done and almost 20% of the respondents often carry the additional work home. Nearly 15% respondents ignore their health and do not engage into any kind of exercises to take care of their health. However, 72% respondents admit that they thoroughly enjoy doing their work. Respondents expressed that they felt a deep sense of satisfaction by contributing to their family and the society at large by being in business and this feeling motivated them to push harder to achieve their goals.

Narratives from Selected Respondents:

During the early months of 2018, the respondent was muddling with the data collection after making all possible efforts to contact the right kind of respondents, the researcher reached out to 158 respondents across the four cities of Ahmedabad, Rajkot Surat and Vadodara. The city-wise break up of respondents is as under:

S. No.	City	No. of Respondents
1.	Ahmedabad	50
2.	Rajkot	19
3.	Surat	50
4.	Vadodara	39
TOTAL		**158**

Each respondent had a unique experience to share and almost everyone had a journey full of struggles and setbacks. But what made them what they are today is their perseverance and patience and the undying will to never give up. The researcher has tried to incorporate few narratives from the respondents as part of the study to show how women work in different trades and enterprises. The researcher has taken care of the ethical aspects and considerations before producing the case studies. Owing to these ethical considerations, the names of the respondents have not been disclosed for the purpose of confidentiality. Thus, narratives have been framed to display the issues and challenges faced by the respondents. The researcher has also attempted at trying to correlate the observations and notes along with the Social Work Skills to identify how these skills have been used by the respondents, and to what extent. Despite the pool of data available, the researcher has limited to sharing one case study per zone to justify the collected data as well as to show variety in the approach of respondents within the same state.

Making the best out of waste

1. **Name of the Organization:** WastePro, Ahmedabad
2. **Year of Establishment:** 2011
3. **Objective:** Understand, Update, Upgrade and undertake all steps possible in biodegradable waste treatment and management.
4. **Activities:** The WastePro team is a team of spirited individuals committed to change the way citizens view waste. To ensure their objectives are met, the team empower children and citizens to become change agents themselves. They also bring the Do It Yourself (DIY) concept in Waste Engagement. Moreover, by this, contribute significantly to Swachh Bharat Mission.
5. **Entrepreneurial Journey:** "I was fed up of the inaction of the municipal corporation & watching cows feeding on plastic waste in their search for food. That is when it first struck me to begin a movement to address this menace and turn into a social entrepreneur," says the respondent.

   The respondent founded her for-profit enterprise in the year 2011, after a career as an Employee for 9 years. During her employment tenure, she worked for credible organizations like ITC Ltd, Confederation of Indian Industry (CII), Autonomous University of Madrid, Spain, Mission Mangalam of the Commissionerate of Rural Development, Govt. of Gujarat and lastly the Gujarat Chamber of Commerce & Industry. The burning desire to turn an employer and start her company led the respondent to tread the uncharted path of a new venture creation specialising in Solid Waste Management. Rearing to venture a male-dominated and government-centric sector was not an easy decision for her as a woman entrepreneur. But, today as an official stakeholder, she has treated tonnes of waste and prevented waste.

   The business model of the respondent is that of treating waste in a decentralized manner, at the point of generation itself. As the project treats a unit's waste in their own premises, garden, food & paper waste is prevented from reaching landfill sites. The venture promotes decentralization of solid waste treatment and management. The enterprise deals with the sales of waste handling tools, vermicomposting tools, earthworms and bio culture.

6. **Beneficiaries:** The respondent has taken her ideas to generate revenues by selling Bioculture, Kitchen waste composting kits, selling earthworms & selling vermi-compost to bungalows, farmhouses, industries & organic farmers. Karnavati Dental School, Mayur Greens Farm House, Shri Uma Industries, Naroda and J K Lakshmi Cement Ltd., Kalol are few of the many beneficiaries.

7. **Issues and Challenges:** The respondent went home to home teaching residents to handle their waste. She has not only toiled to make Ahmedabad a greener city but she has also taught people to derive wealth from waste. She has enlightened many school going children on environment conservation and has transformed lives of many people.

   The biggest challenge that the respondent and her enterprise face today is shifting human consciousness. She believes that it is not the planet that needs saving, it is people. She believes if people start taking environment for granted, there will soon be a time when human race might just be wiped off from the face of the Earth.

8. **Correlation with Dependent Variables:** The respondent's future aspiration is to be known as one of the key social venture companies instrumental in laying the foundation for a Zero Waste Amdavad by 2030.

   The respondent works individually managing her finances and operations with quite a balance. She has a team of workers and labourers not co-workers. She finds it difficult to balance her work-life and personal life; hence, she chooses to live a life by integrating work and family thereby giving her a flexible social life.

9. **Social Work Skills Used (as perceived by the researcher)**: The researcher observed the skills of information gathering and analysis in the respondent. The researcher also observed that the respondent was assertive and direct in her approach. She considers herself as a one-person army and prefers working solo. She is good at problem solving. She has good skills of delegating the work and is inspirational to people to become more environment-conscious.

10. **Learnings:** The respondent prefers to be called a Social entrepreneur. She decided to quit her well-paying job and invest all her savings as seed money for her venture. She believes that nature does not need human beings rather human beings need nature. She has won the best stall award at Vibrant Gujarat 2013. She was a part of the stakeholder's meeting in drafting the new solid waste management rules 2013 of the Ahmedabad Municipal Corporation.

Create your own destiny!

1. **Name of the Organization:** Krish Consultancy, Rajkot
2. **Year of Establishment:** 2006
3. **Objective:** To be the Most Preferred Training and Placement Provider at India Level.
4. **Activities:** The enterprise began with two services - Job Placement (Recruitment and Staffing Solutions) and .HR & Training Implementation Services for Industries and Corporate Houses.
5. **Entrepreneurial Journey:** After Graduating, the respondent decided to start Placement Consultancy firm as she observed that people are willing to work but they do not get the chance or do not have required skills and appropriate information. So in 2006, the respondent started her own consultancy which has now grown to 25 associates across the state, 100 clients and a data base of 1 lac candidates. "Women are generally perceived as home makers with little to do with economy or commerce. However, this picture is changing. In Modern India, more and more women are taking up entrepreneurial activity especially in medium and small scale enterprises", says the respondent. She provides customized Human Resource Management and Training solutions through project based arrangements or complete outsourcing of the human resource function. She has vast experience to identify capability & skill of individual to develop their job career. She is also associated with various Management Institutes and Schools as a Resource Person and Consultant.

   She is passionate to provide best job and career counselling facilities to every job seeker of Saurashtra region through her company. She has served more than 100 Industries & corporate as a HR & Training Consultant. She is a youth icon for youngsters of Saurashtra region, especially the females.

6. **Beneficiaries:** The consultancy has grown over the years and it has been pivotal in providing jobs to over 1 lac candidates within the Saurashtra region. She has provided trainings in almost 100 industries and corporate houses. She has trained more than 10,000 Executives/supervisors in different industries & more than 50,000 students of Institutions.
7. **Issues and Challenges:** She feels that in the region where she belongs, the parents' goal is to educate the females to ensure they get the right match. This mind set hinders many women who have entrepreneurial ideas. Thus, the region sees less contribution of women in entrepreneurial activities as compared to cities like Ahmedabad and Surat.

8. **Correlation with Dependent Variables:** The respondent has contributed in growth & development of SMEs in last 13 years as 5S Expert. She is self-motivated and found space for

her venture because there was a need for such establishment in the market. She is a corporate trainer and leads others be example. She has a dedicated team of trainers and placement providers who work in cohesion to provide employment opportunities to the unemployed youth. Her enterprise follows the model of providing skill-based jobs. Since the enterprise has been long establishes, the respondent finds it very easy to strike balance between her work life and her family life. .

9. **Social Work Skills Used (as perceived by the researcher)**: The researcher observed that the respondent possessed excellent skills of communication and presentation. She possesses right knowledge and owes to her team to keep her updated on the latest news and trends that help her in her training sessions. She is resourceful and has a well-knit network of professionals who push her to fame. She possesses great leadership skills and treats her workers as family. She is empathetic and assertive. She does not believe in jeopardising personal relations at the cost if professional negligence and hence takes utmost care in dealing with her employees. She possess good interpersonal skills and problem-solving skills. She maintains a calm and pleasant disposition.
10. **Learnings:** The respondent believes that life is all about creating yourself so one must design his or her own destiny. The researcher learned that it takes years of struggle to be able to create a space for oneself in the market. The researcher observed this undying spirit to succeed at all costs, in the respondent.

Inspiring people to live WAKEFULLY!

1. **Name of the Organization:** Happy Minds, Surat
2. **Year of Establishment: 2012**
3. **Objective:** The sole mission and objective of the company is to transform lives.
4. **Activities:** HappyMinds is a multi – faceted organization that is committed to bringing this change by empowering people coming from different walks of life. HappyMinds deals in following three verticals:

   ProMinda: These Training and Workshops aim at mentoring & training individuals & organizations to enable them to reach a higher level of thinking thereby giving a meaning to their existence.

   Leadership AcademyThe Leadership Course will take you from existence to excellence! To create an impact and change the approach from traditional bookish learning to global age practical learning.

   Social Initiative-Jazba: This vertical is driven by the Jazbaa to make a difference to society in unique ways. Sessions on iLearn to iLead, Nurturing Young Trainers, Positivity Drive, etc. are the core fundamentals of this vertical.
5. **Entrepreneurial Journey:** The respondent pursued her Master's Degree in Business Administration from MIT, Pune. There she chose a module which, allowed her to work while studying and got herself placed in a Corporate training company. Her work profile was in the marketing sector but she stretched her working hours to learn as much as possible. She spent her break-times in seminars, where one of her senior used to deliver sessions on 'Life Skills'. She happily served tea and snacks to the participants just to be present there and simultaneously, learn. Even after, tiring working hours, she used to manage her time and make modules from what she learnt in the whole day. Since the beginning, she has followed the mantra – "Wake up, Dress up and Show up". And keeping the same mantra as her motivation, she soon started off as a trainer; however, small the number of audience, she had her mother and her best friend, as her constant. There were times, when people tried to pull her down. They mocked her as some 'modern-day-baba' and went on to call her a 'fanatic'. But she had a clear positive vision and a huge faith in her own work. She had built a strong inner system, relying on which she kept going on and did not let criticism disturb her pace. She then studied NLP (Neuro-Linguistic Program) where the roots of her ideology were nurtured. She then invested herself completely in understanding the thought patterns of the human mind.

6. **Beneficiaries:** HappyMinds offers services like training, seminars and workshop sessions and events on positive parenting, entrepreneurial empowerment, confidence building, communication, homemakers' empowerment, corporate training, youth conferences, positive thinking sessions and many more. The beneficiaries range from the ages of 4+ to 60+ years.
7. **Issues and Challenges:** From going door to door to share her ideas, to giving free seminars and training, the respondent dedicated her initial entrepreneurial years in spreading awareness on mind power and building a strong base of trust and understanding among her growing audience. After almost, half-a-decade of persistent efforts and hard work, she brought the wave of change she always intended to. Now, when she announces her public seminars on either 'Positive Thinking', 'Happiness Therapy', 'Mindpower' or any other Life skill subjects.

   The respondent believes that even today, there is not much awareness on the importance of mental wellbeing. She believes a lot needs to be done and she needs to expand the horizons of her venture to different parts of the country and the world at large.
8. **Correlation with Dependent Variables:** The respondent at the age of 19, dreamt and planned out a certain kind of life for herself and did every possible thing to make it work. However, fate had other plans altogether, none of it worked out the way she wanted. Broken and shattered, she started once again. This time her mind was railing with the idea of understanding how it all works. She intended to know how the mind over matter changed the game. That is precisely the time when things started playing out in her favour. This shows the level of motivation of the respondent. She is highly motivated by her own success and aspires to do more to make the world a better place. From a trainer, she has now become an influencer.

   The organisation has a flat model, having more than 25 volunteers associated with them. The respondent, who is the boss, chooses to be called Director and Chief Mentor while she prefers to call her staff members as Chief Energy Officers (CEOs). She takes fun seriously and ensures all employees are always happy and cheerful. She follows a leadership model that further empowers the staff and volunteers to stay committed to the cause of transformation.

   At an early age of 24 years, the respondent was the youngest speaker on the panel across the country for her Social Contribution at the TEDx TALK platform. BRITT India and iHeart Global Inspiration Women'17 have honoured her as one of the top 60 women entrepreneurs in India with innovative business ideas in 2016.

   The respondent owes her success to her extremely supportive mother, her loving husband and both her families. Moreover, the positivity that her professional life gives her makes her personal life bloom too. Hence, she does not face any significant work-life balance issues.

9. **Social Work Skills Used (as perceived by the researcher)**: The respondent is an extremely confident individual with a very clear understanding of herself and her work. She lives a purpose-driven life and chooses to transform people's lives through her efforts. From the perspective of social work skills, the researcher has observed that the respondent possess excellent skills of communication and presentation. She has great resource mobilization skills and very sound interpersonal as well as intrapersonal skills. She follows the POSDCORB model thoroughly and is an efficient manager. She possesses good problem solving skills. She is accepting of herself and the people around her and believes in holding a non-judgmental attitude towards others. She is an epitome of positivity, enthusiasm and energy.
10. **Learnings:** The researcher learnt about a new area of entrepreneurship, which dealt with infusing leadership among children as young as 4 years. On her visit to the enterprise, the researcher observed how intricately confidence boosting and leadership training is provided to the young minds. The researcher also observed how patiently and calmly the staff engaged with children in their fun activities. It was surprising to see that many parents are actually interested in the leadership camps for children and are willing to pay any amount just to enrol into the HappyMinds Academy.

Serving bowlful of happiness

1. **Name of the Organization:** Messy Bowl (earlier Maggewala), Vadodara
2. **Year of Establishment:** 2013
3. **Objective:** To pioneer in delivering hand-crafted food to people and savor their taste buds.
4. **Activities:** The USP is to create usual instant noodles with a twist and offer variety of flavours for all age groups. Their dishes are served at competitive and affordable prices. Core reason of our good standing in the market is our quality that has authentic flavours. It helps them become honest & trust-worthy brand among Maggi-lovers.
5. **Entrepreneurial Journey:** The respondent ventured into food industry without any prior experience. The respondent was working at the Vadodara Municipal Corporation making a sound living with a regular paycheque rolling in every month. Her life was near to perfect. She was also married to the person of her choice, which made her feel that her life was complete. However, life had a different take on her perceived happiness. She soon realised she was in to a troubled marriage and wanted to help her husband by running her venture. However, her husband belonged to a family conditioned to believe that a stable job is what one needs and abhorred the aspect of entrepreneurship. Moreover, the husband was not happy in letting his wife work while he had no income. However, the respondent stood against all odds to make sure her venture succeeded. Her family and friends immensely supported her and she truly believes that her legal separation from her life partner was like receiving a new life. She is an example of what a woman, who has only self-belief by her side, can do. With her vision to serve best in class quality food at affordable prices, she started a brand, that is today, loved by people of all age groups.
6. **Beneficiaries:** From one outlet in the heart of the cultural capital of Gujarat (Vadodara), it became one of the fastest growing chain in the fast-food industry.

   She has created a pool of happy customers in various parts of the country. Besides having their outlets in Vadodara, the venture has also expanded into cities of Rajkot, Surat, Hyderabad and Chennai.
7. **Issues and Challenges:** Founded in 2013 by a first generation entrepreneur with a belief in pioneering in delivering handcrafted food to people and savour the taste buds. Core reason of their strong foothold in the market, despite selling varieties of only one product, is their quality that has authentic flavours. This strategy helps them become honest & trust-worthy brand among Maggi-lovers. "We do not have a fancy, elaborate menu but whatever we serve is cooked with all our heart", says the respondent.

They faced a big challenge when, in 2015, the Nestle Group landed under a legal scanner being accused of being unsafe and hazardous for consumption. As against the maximum limit of 2.50 parts per million (ppm), the amount of lead detected in the Maggi samples was perilously high at 17.2 ppm. The second offence was mentioning 'No added MSG' on the packaging, which is an act of mislabelling. On June 5, 2015, a nationwide ban was imposed on Maggi by FSSAI (Food Safety and Standards Authority of India). During this phase, the respondent had to go out of business for a while s there was a public outrage on serving a banned product. The respondent had to change the menu at her restaurant chain and had to wait until the matter cleared. That was a time when the respondent's outlet was named Maggewala. Once Nestle re-established itself back in the market, the respondent changed the name of her outlet from Maggewala to Messy Bowl. She continues to serve the same product with guaranteed safety. The product and the food chain gained such popularity that she was invited as a guest at the TV Show - Maggi Kitchen Journeys, hosted by the renowned Bollywood actress Smt. Renuka Shahane Rana (______________________________).

8. **Correlation with Dependent Variables:** The respondent kept a very low profile and was happy with her life. After her marriage and shift to Ahmedabad, she came across many street vendors who were experimenting and serving Maggi noodles. On one such evening, when she was out with her friends sharing a bowl of Maggi, it occurred to her to leverage on this idea and open a chain in her hometown – Vadodara. Things seemed to be falling in place, until she found out that her husband did not have a job and she stepped forward to help with her venture. She was shocked to learn that the husband was not in favour of her working and restricted her movement. It was then that her mother stepped forward and helped her come out of a failed marriage. Leaving this behind, the respondent felt more confident in taking the leap and was motivated to devote more hours to her venture.

   The respondent was supported immensely by her family and few friends who helped her choose the right location, staff, furniture, marketing, advertising, publicity and streamlining things for her. She believes in bringing people together and leads by example. She feels if one provides proper attention to the details, it can win many hearts. She does not believe in authority and treats her staff as family.

   Since initially the venture required more involvement, the respondent had to spend days and nights at the restaurant ensuring that everything was in order. At that time, it was difficult to manage her work life and her family life. Gradually, when things started working out favourably, she was able to take time off for her personal commitments. Today, she has a well-balanced life at both personal and professional fronts.

9. **Social Work Skills Used (as perceived by the researcher)**: The researcher observed that the respondent possessed great managerial skills. She had a vision and was able to persevere through all the troubles to make sure her vision turned to reality. She applies the POSDCORB model very efficiently. She displays excellent interpersonal and intrapersonal skills. She possesses sound leadership skills and good skills of resource mobilization. She thrives on her networking skills, which have led to the nationwide expansion of her outlets. She has great problem solving skills. She maintains a pleasant disposition and is always ready to continue to inspire young and potential entrepreneurs with her self-belief and positivity.
10. **Learnings:** The researcher learned that eureka moment could also occur during the times one least expects it. In case of the respondent, it occurred over a bowl of Maggi. Today, that has turned into a successful business venture. The researcher learned that entrepreneurial success does not come overnight and entrepreneurs have to persevere hard to ensure that their vision becomes a reality. The researcher also learnt that despite the difficulties, one needs to pursue and put their foot down to create great results.

SUGGESTIONS & RECOMMENDATIONS

One of the objectives of the study was also to suggest a framework to promote women entrepreneurship in Gujarat. A framework for promoting entrepreneurship among women in Gujarat will be duly discussed and suggested in the next chapter.

This chapter presents the suggestions of the researcher and the recommendations along with areas of Social Work Intervention based on the findings of the analysis elaborated in the previous chapters.

This chapter is further divided in to two parts:

1. Suggestions & Recommendations

2. Areas of Social Work Interventions

1. Suggestions and Recommendations

One of the objectives for this study was to also provide a framework to develop entrepreneurship among women in Gujarat. From the study undertaken, the researcher has got many new insights about female entrepreneurship which were not known earlier. However, there are certain loopholes, to cover up which, the researcher provides suggestions as part of the study.

It is said that an entrepreneur is someone who possesses a risk taking ability and the courage to face problems and suggest alternate and novel suggestions for the same. In case of women entrepreneurs, the expectation is different because if their risk fails, they have no other choice but to shun their dreams of enterprising. It is with this perspective that the suggestions have been provided.

a. Promoting positive attitude towards entrepreneurship through role models

One of the biggest reasons women remain laid back in the field of entrepreneurship is the negative conditioning towards failure and a deep understanding that risk taking almost always leads to failure. This negativity can only be countered with a positive infusion of successful women entrepreneurs taking the lead in coaching the nascent women entrepreneurs.

- Enterprising women are different from small scale vendors and hence their psychosocial outlook is very different. Their education and skills can be effectively used for creating proper forums for budding entrepreneurs wherein the stable entrepreneurs can explain the challenges and mentor the budding entrepreneurs to nurture better.
- The dedicated forums for women entrepreneurs can offer periodical trainings which can cater and invite more and more participation from enterprising women in order to hone their entrepreneurial skills.
- Voluntary role-model engagement programmes can be encouraged where successful women entrepreneurs come together at public events to encourage entrepreneurship among women.
- These forums can be utilised to incubate budding entrepreneurs wherein the skill development can be rigorously emphasised.
- These forums can also act as a platform for women entrepreneurs to effectively conceptualise their entrepreneurial ideas and convert it in to successful businesses.

b. Approaching the proper authorities

- There is an increase in the emphasis of promoting women entrepreneurship under the aegis of Skill India Program of the Government of India and the state government as well. However, there are many entry points for women and many times this creates a confusion among those who are actually seeking help.
- It is suggested to have a single-window system for the registration to loans to scholarships and benefitting schemes, and other aspects related to entrepreneurship. Here, it becomes important to understand that women entrepreneurs may not always have a self-help group (SHGs) or support groups, they may be individuals willing to start up a firm on their own. For such women, a separate system must be devised without them having to run from pillar to post to ensure their paperwork is in place.
- Use of the internet and online systems can prove to be a great boon to the government functionaries and the beneficiaries alike in order to streamline processes for every aspect of entrepreneurship.

- Enterprises registered by women in the name of women must be specially recognised and some rewards can be provided to those enterprises. This strategy may encourage other women also to come up with novel ideas and can lead to increased entrepreneurship.

c. Facilitate access to finance through financial literacy and a range of financing instruments

- There are various financial schemes available for women entrepreneurs in India but the beneficiaries are unaware about it because e of lack of awareness and proper advertisement.
- It is observed that women entrepreneurs lack financial literacy and that financial institutions are also not very open about the explanation regarding the complicated policies.
- Modern methods like "fintech" as an industry sector is changing the nature of business funding, and small businesses are increasingly using a wider range of sources of capital.
- Use of Crowdfunding is also becoming more popular and is applied in many forms, including rewards or gifts, debt, equity, or a combination of debt and equity. Crowdfunding assures of reaching out to a large network of potential financiers over the internet without any face-to-face contact. By resorting to these modern methods, female entrepreneurs may be able to overcome the traditional discrimination in accessing financial backing to start a new ventures. (Halabisky, Potter, Garcia, & Levie, 2016)

d. Easing journey from the mind to the market

- Entrepreneurship is all about the right ideas at the right time combined with the right resources and we have a successful enterprise. However, when it comes to enterprising women, a lot of ideation remains limited only within the enclosures of their minds. The scepticism of failing does not let women take many risks. Moreover, the linkages between the production and marketing of the same are found to be very weak and in their nascent stage.
- Inability to promote their own products or services leaves women dejected and withdrawn from entrepreneurship. This can be explained as every woman knows how to cook, but not every woman is a restaurateur.
- In Gujarat, a government arm of the Mission Mangalam - Gujarat Livelihoods Promotion Company (GLPC) Limited has initiated exhibitions where women entrepreneurs can rent out stalls to market their products. However, these kind of initiatives need to be increased and promoted at a higher level so as to cater to a larger population.

e. Promoting Work life Balance and Family Counselling

- The social roles provoke women to ensure the house is kept in proper order while the man is at work. However, this cannot be convincingly said in case if the women begin working and leave the house for the significant other members to take care of. Hence, sensitization within families needs to be created so that they can promote women in their entrepreneurial activities.
- Young female entrepreneurs also have the responsibilities of marriage and family due to which their entrepreneurial role takes a back seat. A supportive and understanding family can help women overcome their inhibition even during these life events.
- Sensitization to explain that women can also portray the role of an entrepreneur must be created right from the beginning.
- Schools and colleges must adopt Entrepreneurship as part of their curriculum so that idea generation can start at a very young age irrespective of the gender and that deeper family understanding is created to ensure proper support.

2. Areas of Social Work Interventions

The accepted Global definition of Social Work Profession defines that "Social work is a practice-based profession and an academic discipline that promotes social change and development, social cohesion, and the empowerment and liberation of people. Principles of social justice, human rights, collective responsibility and respect for diversities are central to social work. Underpinned by theories of social work, social sciences, humanities and indigenous knowledges, social work engages people and structures to address life challenges and enhance wellbeing." (International Federation of Social Workers, 2014)

Hence, the professional social work respects diversity and enhances well being by use of various methods by which an individual, group or community at large can be empowered to help itself even in absence of any individual or situation.

The gamut of social work has always focused its attention on the weaker and vulnerable sections of the society and tried to mainstream them by countering the difficulties and enhancing welfare.

Through the current study also, the researcher has tried to bring about interventions which can augment entrepreneurship among women and help them be self-reliant.

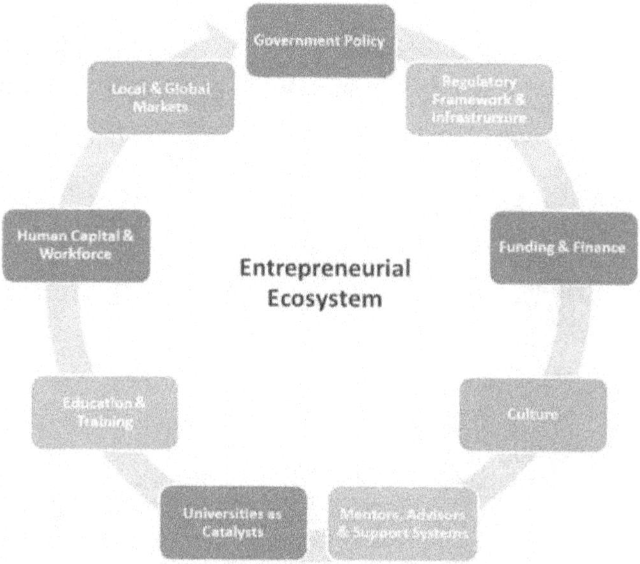

Figure 6.2.1 Showing Entrepreneurial Ecosystem

Figure 6.2.1 briefly explains the entrepreneurial ecosystem in which there are a lot of factors which play their role on helping the entrepreneur rise above their inhibitions and become a successful business-owner. Each of these elements of the entrepreneurial ecosystem can be linked with Social Work Methods to prepare a proper framework for intervention.

Social Work has two methods:

1. Primary Methods

    1.1. Social Casework

    1.2. Social Groupwork

    1.3. Community Organization

2. Secondary Methods

    2.1 Social Welfare Administration

    2.2 Social Work Research

    2.3 Social Action

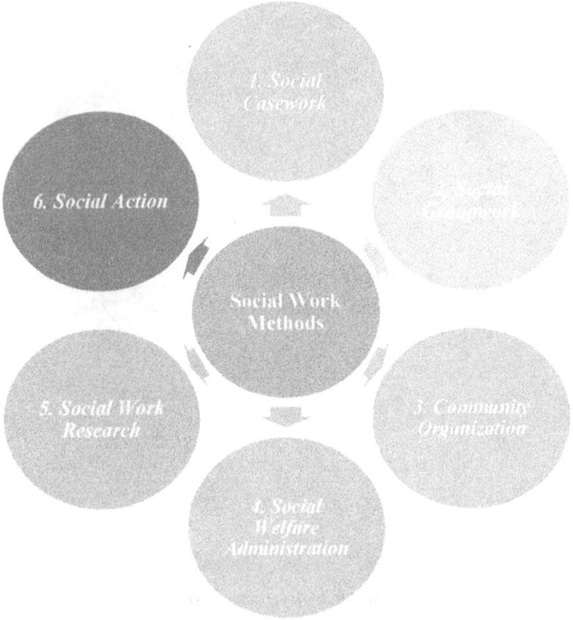

Each method is unique in its approach to ease out human lives by providing scope for improvement at every aspect. Each aspect is described in detail below with areas for Social Work intervention at various levels to augment and accelerate the growth of women entrepreneurship in the state and the country at large.

1. Social Casework

Mary Richmond (1917) defined Social Casework as "the art of bringing better adjustment in social relationships of men, women or children." The purview of casework is dealing with individual problems and help them overcome it.

By applying the theories of casework to the current study, women can be empowered with greater confidence within themselves. In the Indian context, family dependency and personal obligations on women create hindrances to them in succeeding in their career. Motivational factors: Factors like intrinsic motivation, government policies, financial assistance and family support are still in weak shape in India for women entrepreneurs. These individual issues can be taken up and resolved in order to facilitate women. Providing proper hand-holding until these women are fully confident in the way they carry themselves in the market.

2. Social Groupwork

Trecker (1955) defined Social Groupwork as "a method through which individuals in groups and in agency settings are helped by a worker who guides their interaction in programme activities so that they may relate themselves to others and experience growth opportunities in accordance with their needs and capacities to the end.

With the help of the method of Social Groupwork, a focused approach can be followed to work in the direction of empowering women. Continuous and rigorous trainings can be provided to a small group of budding women entrepreneurs and helping them understand group processes and ways to handle their employees and practice leadership. With constant and consistent efforts in training and skilling the women entrepreneurs, they can be made to become more confident as a group.

Moreover, regular meetings with groups of women emphasizing on experience sharing and testimonies can help other women gain inspiration from the already successful women.

3. Community Organization

Ross (1967) defines community organisation as a "process by which a community identifies its needs or objectives, gives priority to them, develops confidence and will to work at them, finds resources (internal and external) to deal with them, and in doing so, extends and develops cooperative and collaborative attitudes and practices in the community".

A reference-network of women entrepreneurs also can be considered as one community from the many communities of the society. This community can be made self-sustaining by developing confidence in them, creating awareness of the various schemes which can provide financial assistance and organizing training programs for them.

Moreover, the various methods of community organization like Education promotion to promote entrepreneurial education, Legislative promotion to promote various schemes offered by the government organizations, non-legislative promotion to promote the schemes offered by voluntary organizations, fund raising for enterprises, conferences of reference networks and evaluation of the status of enterprises can also be done for the smoot functioning of enterprises.

4. Social Welfare Administration

Harleigh Tracker (1971) interprets social welfare administration as a "process of working with people in ways that release and relate their energies so that they use all available resources to accomplish the purpose of providing needed community services and programmes." (Devi & Ravi, 1998)

The government agencies and private agencies which exist in providing support to the enterprising women can be made more active in their functions so that a regular follow-up on trainings is taken and need-based programmes are designed for the benefit of the women entrepreneurs. The District Industries Centers (DICs), the Centres for Entrepreneurship Development, Chambers of Commerce and Industries, national and cooperative banks and all other significant agencies should work in collaboration in order to properly assist women in their trade and business guidance. A single-window agency system must be implemented wherein women can get their issued resolved under one single roof.

5. Social Work Research

Social work research may be defined as systematic investigation into the problems in the field of social work. (Das, 2000)

There is a great scope of researching in the field of women entrepreneurship from a social work perspective. The researcher has made a humble attempt in identifying and studying certain psychosocial issues of entrepreneurship among women but many areas such as impact of government schemes, impact of financial literacy, market linkages and women, leadership, gender analysis of entrepreneurship can be undertaken to gather insights from the field and can be researched upon.

Research in any topic will yield to generation of further scope of study. Similarly, the current study though has limitations, yet many researches can be formulated on the lines of the same topic as that under current study.

6. Social Action

Singh (1986) explains that social action is a process in which conscious, systematic and organised efforts are made by some elites and/or people themselves to bring about change in the system which is instrumental in solving problems and improving conditions which limit the social functioning of weaker and vulnerable sections. (Mishra P. D., 1992)

It is only by collaborating the various efforts made by the various agencies that something positive in the direction of helping women entrepreneurs grow and develop well can be achieved. Proper use of advocacy of the rights of women and women entrepreneurs can help foster the area of entrepreneurship among women. Models of Locality Development and Social Planning can be effective in practicing Social Action within the entrepreneur community.

Thus by working on these aspects, a proper intervention strategy for empowering women through entrepreneurship can be made.

From the above, it can be summarized diagrammatically that development of women entrepreneurship basically requires certain common and collective actions to be undertaken. If designed and implemented well, these actions can pave a new road towards achieving and equal society for women entrepreneurs. These actions are:

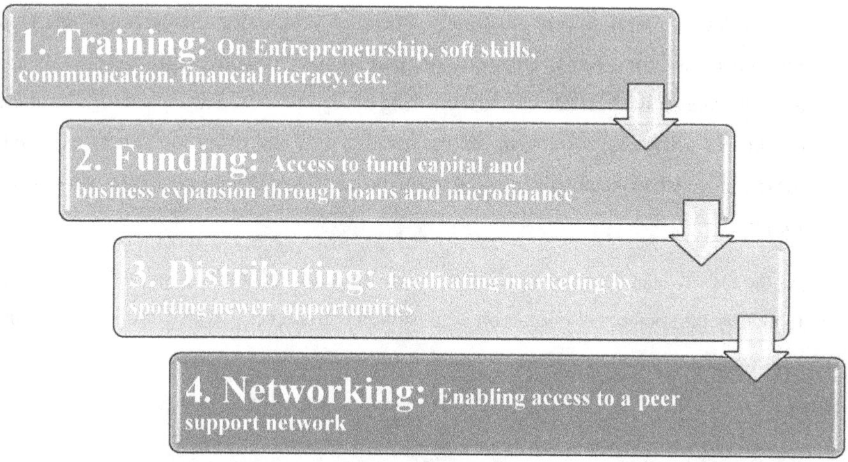

1. Training: Effective trainings can be planned and arranged for focused groups of women entrepreneurs wherein they can be trained on various entrepreneurial aspects like Proposal Formation, Soft Skills, Communication, Financial Literacy, Banking Procedures, Government

Schemes available for support, collaborating and networking as well as overall entrepreneurship development. A proper handholding mechanism must be designed which helps women overcome their diffidence.

2. Funding: Proper linkages must be created between financing agencies and the beneficiaries. Technology and crowdfunding can also be used to its fullest potential to encourage women entrepreneurs. Understanding the need of the business and learning to respond to it by expanding the capital investment with the help of loans and other means of financial support must be encouraged among women. Awareness of these facilities must be spread through proper media so that the number of beneficiaries may rise.

3. Distributing: Effective market linkages must be created in order to facilitate and encourage greater participation from women entrepreneurs. Due to lack of avenues to market their products and services, many women give up on their entrepreneurial dream. This can be avoided if a facilitating market connect can be established with the help of appropriate authorities. Internet has revolutionised the way businesses work these days. Understanding the using digital platforms and social media for marketing can be introduced so that women entrepreneurs can easily market their products online.

4. Networking: Businesses are known to flourish if one has the right kind of contacts. Hence networking becomes a very important aspect of developing entrepreneurship. Dedicated platforms and forums can be created where budding women entrepreneurs can easily approach and get proper guidance, advice and mentoring on how to grow and expand their business. This can enable a learning ecosystem for the beneficiaries and every individual, experienced or inexperienced in managing their business will be benefitted thereby creating a win-win situation for all.

As the society is undergoing a major paradigm shift towards technology, there is requirement to modify the society beliefs and provide women entrepreneurs with a strong ecosystem for entrepreneurship and democracy.

On a concluding note, the researcher would like to quote a very famous poem from one of the very well known and renowned Afro-American Women's Rights Activist, Maya Angelou (Angelou, 2011) on women. The focus and context of her work was basically expressing the extraordinary characteristics of women, however, the researcher has taken liberty to change some words in order to make it more meaningful and apt in the context of the current study.

Home makers wonder where my secret lies

Working women question my business size

But when I start to tell them,

They think I'm telling lies.

I say, it's in the reach of my arms,

the span of my waist

the stride of my step

the curl of my hair, in my fingers, interlaced

I'm a Woman. Phenomenally.

Phenomenal Woman. That's Me.

Inspired by
Maya Angelou

Bibliography

A String of Pearls - Women Entrepreneurs of Sri Lanka: A Casebook. (1995). New Delhi: Asian and Pacific Centre for Transfer of Technology.

Akhtar , S., Nagien, S., & Khan, P. (2009). *Economic Development: A Prospective Study of Women's Entrepreneurship.* (B. Kumar, Ed.) New Delhi: Regal Publications.

Allen, D., Herst, L., Bruck, S., & Sutton, M. (2000). Consequences associated with work to family conflict: A review and agenda for future research. *Journal of Occupational Health Psychology, V*(2), 278-308.

Ambedkar, B. R. (1987). Women and Counter Revolution (Riddles of Hindu Women). In B. R. Ambedkar, *Dr. Baba Saheb Ambedkar: Writings and Speeches* (Vol. III). Department of Education, Govt of Maharashtra.

Anderson, R. S., Grant, M. M., & Speck, B. W. (2008). *Technology to Teach Literacy: A Resource for K-8 Teachers.* Michigan: Pearson/Merrill Prentice Hall.

Angelou, M. (2011). *Phenomenal Woman: Four Poems Celebrating Women.* New York: Random House Publishing Group.

Arya, S. (2000). *Women Gender Equality and the State.* New Delhi: Deep and Deep Publications.

Banerjee, T. (2012, March 11). *Here's How The Status Of Women Has Changed In India [Since 1950 Till Date].* Retrieved March 29, 2017, from YouthKiAwaaz: https://www.youthkiawaaz.com/2012/03/heres-how-the-status-of-women-has-changed-in-india-since-1950-till-date/

Bangwal, D., & Tiwari, P. (2015). Green HRM – A way to greening the environment. *IOSR Journal of Business and Management (IOSR-JBM)* , 45-53.

Bansal, R. (2013). *Follow Every Rainbow.* Chennai: Westlan Ltd.

Barnwal, B. K. (2014, Mar-Apr). Dr. B. R.Ambedkar's Quest for Gender Equality It's Relevance in Contemporary Feminist Discourse. *Online International Interdisciplinary Research Journal, Volume-IV*(Issue-II).

Bent-Goodley, T. B. (2002). DEFINING AND CONCEPTUALIZING SOCIAL WORK ENTREPRENEURSHIP. *Journal of Social Work Education*, 291-302.

Bhansali, V. K. (1999). *Entrepreneurship Development in Next Century: Entrepreneuship, STrategic Managemement and Globalization.* Jaipur: Rawat Publications.

Brush, C. G. (2009, September). *Women Entrepreneurs: A Research Overview.* Retrieved from Oxford Handbooks Online:

http://www.oxfordhandbooks.com/view/10.1093/oxfordhb/9780199546992.001.0001/oxfordhb-9780199546992-e-23

Carlson, S. D., Kacmar, M. K., & Williams, J. L. (2000). Construction and Initial Validation of a Multidomensional Measure of Work-family Conflict. *Journal of Vocational Behaviour*, 249-276.

Carr, M. (1990). Women in Small Scale Industries - Some Lessons from Africa. *Small Enterprise Development, I*(I), 47.

Carsrud, A. L., & Brannback, M. E. (2007). *Entrepreneurship.* Westport, Connecticut: Greenwood Publishing Group.

Cassidy, T., & Lynn, R. (1989). A multifactorial approach to achievement motivation: the development of a comprehensive measure. *Journal of Occupational Psychology*, 301–312.

Chakraborty, D., & Barman, R. R. (2014). A Study on Impact of Motivational Factors on the Growth of Women Entrepreneurship in Assam. *IOSR Journal of Business and Management, 16*(03), 71-80. Retrieved May 2017

Charantimath, P. (2014). *Entrepreneurship Development - Small Business Entreprise.* New Delhi: Pearson Education India.

Chaturvedi, A. (2016, April). A STUDY OF GREEN HRM PRACTICES ADOPTED IN THE SELECTED INDUSTRIAL ORGANIZATIONS . Vadodara: Unpublished.

Chaturvedi, R. (2015). *Work-life Balance of Women Workers.* Vadodara: Unpublished Dissertation.

Cherunilam, F. (n.d.). *Business Environment: Text and Cases.*

Chidambaram, K., & Alagappan, V. (1999). *Business Environment.* New Delhi: Vikas Publishing House.

Chitkara, M. G. (2002). *Dr. Ambedkar and Social Justice.* New Delhi: APH Publishing.

Christensen, J. (2015). Acting locally, Thinking Globally in Social Work Education. *Social Work Education in Europe: towards 2025*, (pp. 1-15). Milan.

Clifton, D., & Amran, A. (2011). The Stakeholder Approach: A Sustainability Perspective. *Journal of Business Ethics*(98), 121–136.

Coate, S. J. (2004). *Ecology and Social Work.* New York: Paul & Co.

Cochran, T. C. (1968). Entepreneurship. In D. L. Sills, & R. K. Merton, *International Encyclopedia of the Social Sciences* (p. 87). New York: Macmillan Co. and Free Press.

Collins, C. J., Hanges, P. J., & Locke, E. A. (2004). The Relationship of Achievement Motivation to Entrepreneurial Behavior: A Meta-Analysis. *Human Performance, 17*(1), 95-117.

Collins, C. J., Hanges, P. J., & Locke, E. A. (2004). The Relationship of Achievement Motivation to Entrepreneurial Behavior: A Meta-Analysis. *Human Performance, XVII*(1), 95-117.

Coughlin, J. H., & Thomas, A. R. (2002). *The Rise of Women Entrepreneurs: People, Processes, and Global Trends.* Westport, Connecticut: Quorum Books.

Cox, D., & Pawar, M. (2006). *International Social Work: Issues, Strategies and Programs.* New Delhi: Vistaar Publications.

Crosby, B. C., & Bryson, J. M. (2005). *Leadership for the Common Good: Tackling Public Problems in a Shared-Power World.* San Francisco: Jossey-Bass.

Dale, M. (2012, January). *Social Work Takes on Business.* Retrieved from NASW NEWS: http://www.socialworkers.org/pubs/news/2012/01/social-work-business.asp

Damor, K., & Mishra, P. (2016, April 22). Entrepreneurs check into women-only industrial park. *The Times of India.*

Danto, E. A. (2011, May 25). *Postmodernism and Social Work.* Retrieved from Oxford Bibliographies: http://www.oxfordbibliographies.com/view/document/obo-9780195389678/obo-9780195389678-0132.xml

Das, D. K. (2000). *Practice of Social Research.* Jaipur: Rawat Publications.

Das, D. K. (2000). *Practice of Social Research : A Social Work Perspective.* Jaipur: Rawat Publications.

Das, R. (1998). *The Environment Divide: the Dilemma of Developing Countries.* New Delhi: Indus.

Davis, G. (2010). *Closing Key-note Address.* New York: United Nations Headquarters. Retrieved March 28, 2017

Deepika, R., & Kapargam, V. (2016). A study on green HRM practices in an organisation. *International Journal of Applied Research*, 426-429. Retrieved 2017

Desai, M. (2002). *Ideologies and Social Work: Historical and Contemporary Analyses.* Jaipur: Rawat Publications.

Desai, V. (2007). *Entrepreneurship Development and Management.* Mumbai: Himalaya Publishing House.

Desai, V. (2009). *The Dynamics of Entrepreneurial Development and Management.* New Delhi: Himalaya Publishing House.

Deshmukh, S., & Deshmukh, P. (2016). Improving HR Interventions Related to Work Life Balance for Educational Sector in Indian Scenario. *Intercontinental Journal of Human Resource Research Review*, 165-176.

Devi, R., & Ravi, P. (1998). *Social Work and Social Welfare Administration, Methods and Practices (Vol I).* Jaipur: Mangal Deep Publications.

Dhameja, S. K. (2002). *Women Entrepreneurs: Opportunities, Performance and Problems.* New Delhi: Deep & Deep Publications Pvt. Ltd.

Dhani, S. L. (2007). *Dr. B.R. Ambedkar: Man of Millennium For Social Justice.* New Delhi: Kalpaz Publications.

Dhanvijay, V. (2012, April). Dr. Babasaheb Ambedkar's Efforts for Women Empowerment and Present Status of Women in Society. *Electronic International Interdisciplinary Research Journal, Volume-I*(Issue II).

Dhawan, A., & Dalmia, N. (2015, July 08). The Digital India Push for Education. *The Financial Express*.

Dixit, M. (1998). *Women and Achievement: Dynamics of Participation and Partnership.* New Delhi: Kanishka Publishers, Distributors.

Dobers, P. (2009). Corporate Social Responsibility: Management and Methods. *Corporate Social Responsibility and Environmental Management*, 185-191.

Dodh, P., Singh, S., & Ravita. (2013). Corporate Social Responsibility and Sustainable Development in India. *Global Journal of Management and Business Studies*, 681-688.

Dominelli, L. (1990). *Women and Community Action.* Birmingham: Venture Press.

Dominelli, L. (2005). *Social Work: Theory and Practice for a Changing Profession.* Massachusetts: Polity Press.

Drucker, P. (1973). *Management: Tasks, Responsibilities, Practices.* New York: Harper & Row.

Drucker, P. (2014). *Innovation and Entrepreneurship.* Routledge: New York.

Drucker, P. F. (1986). *Innovation and Entrepreneurship, Practice and Principles.* London: Pan Books.

Dubrin, A. J. (2004). *Leadership: Research Findings, Practice and Skills* (4th ed.). New York: Houghton Mifflin Company.

Eddleston, K. A., & Powell, G. N. (2012). Nurturing Entrepreneurs' Work–Family Balance: A Gendered Perspective . *Entrepreneurship Theory and Practice*, 513-540.

Eraut, M. (1994). *Developing Professional Knowledge and Competence.* London: Falmer Press.

Erickson, C. L. (2018). *Environmental Justice as Social Work Practice.* New York: Oxford University Press.

Feminism. (2015, August 14). Retrieved from UK Essays: https://www.ukessays.com/essays/social-work/feminism.php?cref=1

Fielden, S. L., & Dawe, A. (2004). Entrepreneurship and Social Inclusion. *Women in Management Review, 19*(3), 139-142.

Finn, J. D. (2000). *Extending Education Through Technology: Selected Writings by James D. Finn on Instructional Technology.* (R. McBeth, Ed.) Connecticut: Information Age Publishing.

Flexner, S. B. (1987). *Random House Webster's Unabridged Dictionary.* London: Random House.

Fondas, N. (1997). Feminization unveiled: Management Qualities in Contemporary Writings. *Academy of Management Review*, 257-282.

From Ideas to Independence: A Century of Entrepreneurial Women. (n.d.). Retrieved from National Women's History Museum: http://entrepreneurs.nwhm.org/#/1910-1939/1

Ganesan, S., & Duraipandian, R. (1999). Promoting Entrepreneurship among Rural Women. In M. Soundarapandian, *Women Entrepreneurship: Issues and Strategies* (pp. 62-67). New Delhi: Kanishka Publishers.

Geetha, P., Prema, M., & Jansi, M. (2012). *Women in Management: Leadership and Capacity Building.* New Delhi: Mittal Publications.

Giddens, A. (2001). *The Consequences of Modernity.* London: Polity Press.

Global Definition of Social Work. (2014, August 06). Retrieved from IFSW: http://ifsw.org/policies/definition-of-social-work/

Goffee, R., & Scase, R. (1985). *Women in Charge: The Experiences of Female Entrepreneurs.* London: Unwin Hyman.

Gond, J.-P., & Moon, J. (2011). Corporate Social Responsibility in Retrospect and Prospect:Exploring the Life-Cycle of an Essentially Contested Concept. In R. Caruana, *Routledge Major Work on Corporate Social Responsibility* (pp. 01-40). Nottingham: Nottingham University Business School.

Gray, M., Coates, J., & Hetherington, T. (2012). *Environmental Social Work.* Routledge: New York.

Greaves, T. W. (2012). *Revolutionizing Education Through Technology: The Project RED Roadmap for Transformation.* Virginia: International Society for Technology in Education.

Green, E., & Cohen, L. (1995). 'Women's business': Are women entrepreneurs breaking new ground or simply balancing the demands of 'women's work' in a new way? *Journal of Gender Studies*, 297-314.

Greenhaus, H. J., & Collins, H. K. (n.d.). The Relation between Work Family Balance and Quality of Life. *Journal of Vocational Behaviour*, 510-531.

Greenhaus, H., Collins, H., & Shaw, D. (n.d.). The relation between work family balance and quality of life. *Journal of Vocational Behaviour*, 510-531.

Gretczko, M., & Mazor, A. (2016, March 10). Designing HR for the Moments that Matter. *Human Capital*. Retrieved July 2018, from https://hrtimesblog.com/2016/03/10/designing-hr-for-the-moments-that-matter/

Guenther, J. (Ed.). (2015, September 07). *Girl Child Labourers in India: An Invisible Issue.* Retrieved September 20, 2015, from WikiGender: http://www.wikigender.org/index.php/Girl_Child_Labourers_in_India:_An_Invisible_Issue

Guha, R. (2011). *Makers of Modern India.* Cambridge: Harvard University Press.

Gunjal, V. R. (2012). Dr. Babasaheb Ambedkar and Women Empowerment. *Social Work*, 84-85.

Gurnani, P. S. (2016). *Women Entrepreneurship: Emerging Dimension of Entrepreneurship in India.* New Delhi: Educreation Publishing.

Halabisky, D., Potter, J., Garcia, M. C., & Levie, J. (2016). *Policy Brief on Women's Entrepreneurship.* Luxembourg: Publications Office of the European Union. doi:10.2767/50209

Hall, J., Matos, S., Sheehan, L., & Silvestre, B. (2012). Entrepreneurship and Innovation at the Base of the Pyramid: A Recipe for Inclusive Growth or Social Exclusion? *Journal of Management Studies, 49*(4), 785–812.

Halme, M., Roome, N., & Dobers, P. (2009). Corporate responsibility: Reflections on context and. *Scandinavian Journal of Management*, 1-9.

Harari, Y. N. (2015). *Sapiens: A Brief History of Humankind.* London: Vintage-Penguin Random House.

Hegde, B. (2013). Dr. B. R. Ambedkar: A Ray of Hope for the Downtrodden in India. *International Journal of Social Science & Interdisciplinary Research*, 144-147.

Hisrich, R. (2011). *Entrepreneurship 6/E.* New Delhi: Tata McGraw-Hill.

Hisrich, R. D., Peters, M. P., & Shepherd, D. A. (2005). *Entrepreneurship.* Boston, MA: McGraw-Hill Irwin.

Holmqvist, M. (2009). Corporate social responsibility as corporate social control: The case of work-sitehealth promotion. *Scandinavian Journal of Management*, 68-72.

Hughes, K. (2006). Exploring Motivation and Success Among Canadian Women Entrepreneurs . *Journal of Small Business & Entrepreneurship*, 107-120.

IANS. (2016, May 07). Women run 14 per cent of Indian businesses, most self-financed . *The Economic Times - Panache.*

Ibarra, H., Ely, R. J., & Kolb, D. M. (2013, September). Women Rising: The Unseen Barriers. *Harvard Business Review*.

International Federation of Social Workers. (2014, July). *Global Definition of the Social Work Profession*. Retrieved July 2019, from International Federation of Social Workers: https://www.ifsw.org/what-is-social-work/global-definition-of-social-work/

International Finance Corporation, World Bank Group. (2014). *Micro, Small, and Medium Enterprise Finance Improving Access to Finance for Women-owned Businesses in India.* New Delhi: International Finance Corporation.

Iyer, L. (2007). *Women Entrepreneurs -Challenges and Strategies.* New Delhi. : Friedrick E.Bert Stifting.

Iyer, L., Khanna, T., & Varshney, A. (2011, October 18). Caste and Entrepreneurship in India. *Working Paper 12-028*. Harvard Business School.

Jackson, S. L. (2009). *Research Methods and Statistics: A Critical Thinking Approach.* Belmont: Wadsworth, Cengage Learning.

Jain, A. (2016). Green Human Resource Management Concept In Indian Industries – Present Scenario. *International Journal of Multidisciplinary Allied Research Review and Practices*, 171-181. Retrieved 2017

Jakobsh, D. R. (2004). Barriers to Women's Leadership. (T. Oaks, Ed.) *Encyclopedia of Leadership*, 77-81.

Jifiti, T. (2016, April 20). *Catching Up with the Elderly: Optimizing eCommerce for the Golden Agers*. Retrieved from Jifiti. Be Gifted.: https://gifts.jifiti.com/blog/gifting-data/optimizing-for-the-elderly/

Johnson, A. (2000). *The Blackwell Dictionary of Sociology* (2nd ed.). Oxford: Blackwell.

Juneja, P. (2018). *Human Resource Management (HRM) - Definition and Concept.* Retrieved August 2018, from Management Study Guide: https://www.managementstudyguide.com/processes-in-human-resource-management.htm

Jurkowski, E. T., & Keefe, R. (Eds.). (2013). *Handbook of Public Health Social Work.* New York: Springer Publishing Company.

Kait, K. (2013, December 18). *Articles: Legal Services India.* Retrieved from Legal Services India: http://www.legalservicesindia.com/article/article/dr-b-r-ambedkar%E2%80%99s-role-in-women-empowerment-1611-1.html

Kannan, S. (n.d.). *Vedas and Modern Business Management Practices* . Retrieved from Ved Vikas: http://vedvikas.com/samplechapter.pdf

Kapila, U. (2007). *Indian Economy: Performance and Policies.* New Delhi: Academic Foundation.

Kaplan, M. (2013, December 16). *Ecommerce Merchants Should Not Ignore Older Shoppers.* Retrieved from PracticalEcommerce: http://www.practicalecommerce.com/articles/61984-Ecommerce-Merchants-Should-Not-Ignore-Older-Shoppers

Kaul, R. (2018, June 07). From 167 to 130, India's maternal mortality rate down by 22%. *Hindustan Times.*

Kaushik, D. S. (2009). *Women Entrepreneurs.* Jaipur: Ritu Publications.

Keer, D. (1954). *Dr. Ambedkar: Life and Mission* . Bombay: Popular Prakashan.

Kellner, D. (2002). Theorizing Globalization. *Sociological Theory, 20*(3), 285-305.

Khanka, S. S. (2006). *Entrepreneurial Development.* New Delhi: S.Chand Publications.

Kidwai, N. L. (2015). *30 Women in Power: Their Voices, Their Stories.* Faridabad: Thomson Press India Ltd.

Kirkwood, J., & Tootell, B. (2008). Is entrepreneurship the answer to achieving work–family balance? *Journal of Management & Organization*, 285 - 302 .

Kirschenhofer, F. (2006, January 13-14th). Long-term performance of habitual entrepreneurs – which direction to go? *Paper presented at the Workshop on Firm Exit and Serial Entrepreneurship at the Max Planck Institute*. Jena, Germany.

Knopik, M. S., & Moerer, T. (2008). Women Leaders and Entrepreneurs: Learning from One to Teach the Other. *Journal of Women in Educational Leadership*, 137-153.

Knopik, M. S., & Moerer, T. (2008). Women Leaders and Entrepreneurs: Learning from One to Teach the Other . *Journal of Women in Educational Leadership*, 137-153.

Kothari, C. R. (2004). *Research Methodology: Methods and Techniques* (Second ed.). New Delhi: New Age International.

Kraisonswadi, N. (1989). *Women Executives: A Sociological Study in ROle Effectiveness.* Jaipur: Rawat Publications.

Krishnan, J. (2013). *Breaking Barriers.* Mumbai: Jaico Publishing House.

Kumar, A. (2012). *Entrepreneurship: Creating and Leading an Entrepreneurial Organization.* Noida: Dorling Kindersley (India) Pvt. Ltd.

Kumar, B. (Ed.). (2009). *Women Empowerment and Sustainable Development.* New Delhi: Regal Publications.

Kumar, R. (2007). *Ambedkar and His Writings: A Look for the New Generation.* New Delhi: Gyan Publishing House.

Kumari, S. (2012). Women Entrepreneurship in India. *International Indexed & Referred Research Journal*, 6.

Lal, A. T. (2014, December). *Women Entrepreneurs in India - Over the Years!* Retrieved from SSRN: https://ssrn.com/abstract=2176377

LalithaDevi, U. (1982). *Status and Employment of Women in India.* New Delhi: B.R. Publishing Corporation.

Lavanya, T. (2010). *Women Empowerment Through Entrepreneurship.* New Delhi: New Century Publications.

Laxmi, C. S. (Ed.). (1998). *Development of Women Entrepreneurship in India.* New Delhi: Discover Publishing House.

Lewis, K. V., Henry, C., Gatewood, E., & Watson, J. (Eds.). (2014). *Women's Enterpreneurship in the 21st Century: An International Multi-level Research Analysis.* Massachusetts: Edward Elgar Publishing Inc.

Limaye, C. (1999). *Women: Power and Progress.* New Delhi: B. R. Publishing Corporation.

Littunen, H. (2000). Entrepreneurship and the characteristics of the entrepreneurial personality. *International Journal of Entrepreneurial Behaviour & Research*, 295–309.

Lokhande, D. M. (2007). A Study of Women Entrepreneurship Development in Industries Backward Areas. *Entrepreneurship for Economic Development Conference* (pp. 1-8). Jalna: Jain Printers.

Loney, M. (1983). *COmmunity Against Government: The British Community Development Project 1968-1978 A Study of Government Incompetence.* London: Heinemann.

Long, D. D., Tice, C. J., & Morrison, J. D. (2012). *Macro Social Work Practice: A Strengths Perspective* . Boston: Cengage Learning.

Lussier, R. N., Corman, J., & Kimball, D. (2014). *Entrepreneurial New Venture Skills.* New York: Routledge.

Malhotra, S. (2015, June 07). *Invisible Hands.* Retrieved from Business Today: http://www.businesstoday.in/magazine/features/child-labour-in-india-how-it-being-hidden-from-authorities/story/219448.html

Manfredi, S., & Holliday, M. (2004). *Work-Life Balance: An Audit of Staff Experience at Oxford Brookes University* . Oxon, Great Britain: The Centre for Diversity Policy Research.

Mathew, R. V., & Panchanatham, N. (2011, July). AN EXPLORATORY STUDY ON THE WORK-LIFE BALANCE OF WOMEN ENTREPRENEURS IN SOUTH INDIA. *Asian Academy of Management Journal, 16*(2), 77–105.

Maxwell, A. G., & McDougall, M. (2004). Work-life Balance: Exploring the connection between levels of influences in UK Public Sector. *Public Management Review*, 377-393.

Mays, G. P., & Novick, L. F. (Eds.). (2005). *Public Health Administration: Principles for Population-based Management.* Burlington, Massachusetts: Jones & Bartlett Learning.

McClelland, D. C. (1961). *The Achieving Society.* New York: Free Press.

McClelland, D. C. (1967). Money as motivator: some research insights. *McKinsey Quarterly* , pp. 10–21.

McClelland, D. C., Clark, R. A., Roby, T. B., & Atkinson, J. W. (1958). The Effect of the Need for Achievement on Thematic Apperception. In J. Atkinson, *Motives in Fantasy, Action and Society* (pp. 64-82). Princeton, NJ: Van Norstrand.

McCormack, J. (2007). Globalization and Social Work Education and Practice: Exploring Australian Practitioners' Views. *The Journal of Sociology & Social Welfare*, 123-142.

McKay, R. (2001). Women Entrepreneurs: Moving Beyond Family and Flexibility. *International Journal of Entrepreneurial Behavior & Research*, 148 - 165. Retrieved February 2018

McKenzie, W. (2005). *Multiple Intelligences and Instructional Technology.* Virginia: International Society for Technology in Education.

McKenzie, W. (2005). *Multiple Intelligences and Instructional Technology.* International Society for Technology in Education.

Mehta, B., & Parekh, S. (2014). Women Entrepreneur: A Road to Empowerment. *Strategies for Business Challenges*, 71 - 75.

Mehta, B., & Parekh, S. (2014). Women Entrepreneurs: Leading the Way Ahead. *Journal of Social Welfare and Management, 6*, 35 - 40.

Menon, N. (2015, April 10). *Worker (Census Definition).* Retrieved September 21, 2018, from IES Arthapedia: http://www.arthapedia.in/index.php?title=Wroker_(Census_Definition)

Mishra, & Puri. (2006). *Indian Economy* (24th ed.). New Delhi: Himalaya Publishing House.

Mishra, P. D. (1992). *Social Work - Philosophy and Methods.* New Delhi: Inter India Publications.

Mishra, S. (2009, September 11). *Women Enterpreneurship Development in India.* Retrieved from IndianMBA: http://indianmba.com/Faculty_Column/FC1073/fc1073.html

Moerer-Urdahl, T. (2005). A longitudinal qualitative study of collegiate mentoring experiences in the Nebraska Human Resources Research Foundation. Doctoral dissertation, The University of Nebraska-Lincoln.

Moharana, S., & Dash, C. R. (1996). *Entrepreneurship Development.* Jaipur: RBSA Publishers.

Mondal, P. (2016). *Status of Women in Vedic and Post-Vedic Period.* Retrieved from Your Article Library: http://www.yourarticlelibrary.com/society/status-of-women-in-vedic-and-post-vedic-period/4397/

More, V. G. (2011, August). Dr. B. R. Ambedkar"s Contribution for Women's Rights. *Variorum - Multi- Disciplinary e-Research Journal, 02*(I).

Mundra, S., & Sharma, J. (2013). A Comparative Analysis of the Status of Women Entrepreneurship: India V/S Global. *Research Journal of Social Science & Management*, 147-156.

Murthy, B., Chandrashekhar, M., & Rao, M. G. (1986). Entrepreneurial Process and Promises. *Precision.*

Murthy, N. (2003). The Essence of Leadership. *The Smart Manager*. Retrieved February 2017

Nalini, M. (1993). Caste and Patriarchy: Ambedkar's Insight into the Status of Indian Women. In K. S. Chalam, *Relevance of Ambedkarism in India* (pp. 129-138). Jaipur: Rawat Oublications.

Nambiyar, S. (2017). *Success Stories of Entrepreneurs of the Banyan City.* Ahmedabad: University Granth Nirman Board.

Nathani, K. (2019). *#9 Powerful Quotes from Successful Indian Women Entrepreneurs.* Retrieved May 2019, from Entrepreneur India: https://www.entrepreneur.com/article/298133

Nayyar, P., Sharma, A., Kishtawaria, J., Rana, A., & Vyas, N. (2007). Causes and Constraints by Women Entrepreneurs in Entrepreneurial Process. *Journal of Social Science, 14*(2), 99–102.

Newar, S. (2011, January 24). *Mother –The True Identity of Woman.* Retrieved from Agniveer: http://agniveer.com/mother-woman/

Nieuwenhuizen, C. (2009). *Entrepreneurial Skills.* Lansdowne, Cape Town: Juta and Company Ltd.

Omvedt, G. (2008). *Ambedkar: Towards an Enlightened India.* U.K.: Penguin.

(2005). *Opportunities for Women Entrepreneurship.* New Delhi: National Institute for Industrial Research.

Pandya, R. (2016). *Skill Development and Entrepreneurship in India.* New Delhi: New Century Publications.

Pandya, R. (2016). *Skill Development and Entrepreneurship in India.* New Delhi: New Century Publications.

Pareek, U., & Rao, T. V. (1992). *First Handbook of Psychological and Social Instruments.* New Delhi: Concept Publishing Company.

Park, K. (2017). *Parks Text Book Of Preventive & Social Medicine* (24 ed.). Jabalpur: Banarasidas Bhanot.

Patel, C. (2013, April). A Study of Women Entrepreneurs of Vadodara City. Vadodara.

Patel, J. R. (2005, May). *Growth of Women Entrepreneurship in Baroda.* Vadodara, Gujarat, India.

Patscher, K., Wade, L., & Simon, P. (2010). *Engaging Philanthropy to Promote Gender Equality and Women's Empowerment.* New York: United Nations Headquarters. Retrieved March 28, 2017

Pearce, C. L., Maciariello, J. A., & Yamawaki, H. (2010). *The Drucker Difference: What the World's Greatest Management Thinker Means.* Pennsylvania: McGraw Hill.

PHD CHAMBER OF COMMERCE AND INDUSTRY . (2019, March). *Women Entrepreneurship: Transforming from Domestic Household to Financial Independence.* Retrieved from PHD Research Bureau: https://www.phdcci.in/wp-content/uploads/2019/03/Women-Entrepreneurship-Transforming-from-domestic-household-to-financial-independence.pdf

Radhakrishnan, & Moore. (1957). Vedas and Upaniṣads. In Phillips, *Philip's Encyclopaedia* (p. 3). London: Phillips.

Rajagopalan, R. (2016). *Environmental Studies: From Crisis to Cure.* London: Oxford University Press.

Rajarajeswari, L. (2012). *Women Empowerment - The Condition and Position of Women.* Retrieved from IndianMBA: http://www.indianmba.com/Faculty_Column/FC1435/fc1435.html

Ramamohan, C. V., Samineni, S., & Vemuri, S. (2011). *EMPOWERMENT OF WOMEN THROUGH LEADERSHIP & ENTREPRENEURSHIP.* New Delhi: Himalaya Publishing House.

Ramsay, S., & Boddy, J. (2017). Environmental Social Work: A Concept Analysis. *The British Journal of Social Work, 47*(01), 68-86.

Rani, D. L. (1996). *Women Entrepreneurs.* New Delhi: A.P.H Publishing Corporation.

Rao, B. S. (2007). *Women Empowerment & Planning Process.* Retrieved March 31, 2017, from Content Writer India: www.contentwriter.in/articles/others/women-empowerment.htm

Reddy, P. N. (1991). Problems of Women Entrepreneurs in Goa: A Pilot Study. *Khadi Gramadyog, 37*(4), 157–159.

Richardson, D., & Robinson, V. (Eds.). (1993). *Introducing Women's Studies.* London: Macmillan Press.

Rizvi, A. F., & Gupta, K. L. (2009). Women Entrepreneurship in India – Problems and Prospects. *OORJA Journal of Management and I.T., VII*(2), 35–41.

Roy, R. (2011). *Entrepreneurship.* New Delhi: Oxford University Press.

Roy, T. (2001). WOmen Entrepreneurs in Urban Informal Sector. *TISS Mumbai.*

Rusinko, C. A. (2010). Integrating Sustainability in Higher Education: A Generic Matrix. *International Journal of Sustainability in Higher Education*, 250-259.

Saha, J. (2006). *Management and Organizational Behaviour.* New Delhi: Excel Books India.

Sangolagi, K., & Alagawadi, M. (2016). Women Entrepreneurs . *International Journal of Advancement in Engineering Technology, Management and Applied Science*, 216-222.

Sarada, S. (2004). *Stress of Women Entrepreneurs: A Comparative Study with Women in Selected Groups in the District of Ernakulam.* Kochi: School of Management Sudies, Cochin University of Science and Technology.

Saurabh, S. (n.d.). ISSUES AND CHALLENGES FACED BY WOMEN ENTREPRENEURS AND THEIR TRAINING NEEDS. *International Journal of in Multidisciplinary and Academic Research (SSIJMAR)July-August (ISSN 2278 – 5973), 1*(2).

Schumpeter, J. A. (1955). Economic Theory and Entrepreneurial History. In H. Aitken, *Explorations in Enterprise* (pp. 51-52). Cambridge: Harvard University Press.

Schwartz, E. B. (1973). Can Women Compete in Small Business? *Journal of Small Business Management* , 4.

Sehgal, R. (2005). Social Work and Industry- Dilemma of Partnership. *Contemporary Social Work*, 66-95.

Sellitz, C., Jahoda, M., Deutsch, M., & Cook, S. W. (1951). *Research Methods in Social Relations.* New Tork: Holt, Rinehart and Winston.

Setty, E. D. (1980). Developing Enterpreneurship Among WOmen. *Man and Development, II*(03).

Setty, E. D., & Krishnamoorty, P. (1981). *Strategies for Developing Women Entrepreneurship.* New Delhi: Akanksha Publishing House.

Shah, H. (2013). *Creating and Enabling Environment for WOmen Entrepreneurship in India.* New Delhi: Economic and Social Commission for Asia and the Pacific (ESCAP). Retrieved October 2019, from https://www.unescap.org/sites/default/files/ESCAP-SSWA-Development-Paper_1304_1.pdf

Shane, S., & Venkataraman, S. (2000). The Promise of Entrepreneurship as a Field of Research. *Academy of Management Review*, 217-226.

Sharma, D. D., & Dhameja, S. K. (2010). *Women Entrepreneurs and Enterprise Management.* Chandigarh: Abhishek Publications.

Sharma, J. P. (2009). *Environmental Studies.* New Delhi: Laxmi Publications.

Sharma, K. (2016, May). *14 Percent of Indian Businesses Run By Women Entrepreneurs.* Retrieved October 2017, from Indian Web2: Technology, Web and Start-ups: https://www.indianweb2.com/2016/05/10/14-percent-indian-businesses-run-women-entrepreneurs/

Sharma, M. (2017). Women Entrepreneurs: A Study of Acheivement Motvation, Leadership and Work-life Balance of Women Entrepreneurs of Gujarat. *Concept Paper*. Unpublished.

Sharma, P. (2014). Changing Status of Women-owned Enterprises in IndiaAn. *Global Journal of Finance and Management*, pp. 933-944.

Sharma, S. (2016). *Entrepreneurship Development.* New Delhi: PHI Learning Pvt. Ltd.

Shekhar, N., & Padmadeo, K. B. (2009). *Women Entrepreneurship and Empowerment.* (B. Kumar, Ed.) New Delhi: Regal Publications.

Shelton, L. M. (2006). Female Entrepreneurs, Work–Family Conflict, and Venture Performance: New Insights into the Work–Family Interface. *Journal of Small Business Management*, 285-297. Retrieved February 2018

Sheopuri, A., & Sheopuri, A. (2015). Green Hr Practices in the Changing Workplace. *Business Dimensions*, 13-26.

Siddiqui, S. (2008). *Women Enterpreneurs in Export Trade.* New Delhi: Regal Publications.

Singariya, M. R. (2014). Dr. B. R. Ambedkar and Women Empowerment in India. *Journal of Research in Humanities and Social Science*, 01-04.

Singh, A. (2014). Role of Women Entrepreneurs in India: A SWOT Analysis. *International Journal of Management and International Business Studies*, 231-238.

Soota, A., & Gopalan, S. R. (2016). *Entrepreneurship Simplified: From Idea to IPO.* UK: Penguin.

Soundarapandian, M. (1999). *Women Enterpreneurship: Issues and Strategies.* New Delhi: Kanishka Publishers and Distributors.

Stevenson, M. (2018, March). Employee Engagement to Employee Experience: Why HR Should focus on the Individual? *HR Exchange Network*. Retrieved August 2018

Swarnalatha, C., & Rajalakshmi, S. (2016). *A HAND BOOK ON WORK-LIFE BALANCE.* Solapur: Laxmi Book Publication.

Tang, S. (2010, June). *Work support, work-family enrichment, work demand and work well-being among Chinese employees : A study of mediating and moderating processes.* Retrieved from Digital Commons @ Lingnan University: http://commons.ln.edu.hk/do/search/?q=author_lname%3A%22TANG%22%20author_fname%3A%22Shuwen%22&start=0&context=2433414

Taylor, S. H., & Roberts, R. W. (2013). *Theory And Practice of Community Social Work* . Jaipur: Rawat Publication.

Teasley, D. (n.d.). *Achievement Motivation: Theory & Definition.* Retrieved from Study.com: http://study.com/academy/lesson/achievement-motivation-theory-definition-quiz.html

Thomas, R. M. (2003). *Blending Research Methods in Theses and Disstertations.* California: Sage Publications Inc.

Tiwari, A. (2007). *Women Entrepreneurship and Economic Development.* New Delhi: Sarup & Sons.

Tiwari, A. K. (2016, April 26). Walmart lends a helping hand to women entrepreneurs. *DNA*.

Tiwari, S., & Tiwari, A. (2007). *Entrepreneurship Development in India.* New Delhi: Sarup and Sons.

TNN. (2016, May 03). Women Entrepreneurs Come Together for Better Business. *The Times of India*, p. 1.

Vaidya, S. (2014). *Developing Entrepreneurial Life Skills: Creating and Strengthening Entrepreneurial Culture in Indian Schools.* New Delhi: Springer Science & Business Media.

Valeri, A., Parton, B., & Robb, A. (2014). *Entrepreneurship Education and Training Programs around the World: Dimensions for Success.* Washington, DC: World Bank Publications.

Vasanthagopal, R., & Santha, S. (2008). *Women Entrepreneurship in India.* New Delhi: New Century Publications.

Vinay, D., & Singh, D. (2015). Status and Scope of Women Entrepreneurship. *Universal Journal of Management* , 43-51.

Virtanen, M. (1997). The role of different theories in explaining entrepreneurship. (S. Kunkel, Ed.) *Entrepreneurship: the engine of global economic development.*

Walker, E., Wang, C., & Redmond, J. (2008). Women and work-life balance: is home-based business ownership the solution? *Equal Opportunities International*, 258 - 275. Retrieved December 2017

Warke, N. P., & Narkhede, P. A. (2014). WOMEN EMPOWERMENT: SKILLS ENHANCEMENT THROUGH ENCOURAGING ENTREPRENEURSHIP. *Tactful Management Research Journal*, 70-75.

Women's Entrepreneurship Development (WED) Programme. (n.d.). Retrieved from International Labour Organization: http://www.ilo.org/empent/areas/womens-entrepreneurship-development-wed/lang--en/index.htm